Cultural Tourism in Southern Africa

TOURISM AND CULTURAL CHANGE

Series Editors: Professor Mike Robinson, *Ironbridge International Institute for Cultural Heritage, University of Birmingham, UK* and Dr Alison Phipps, *University of Glasgow, Scotland, UK*

TCC is a series of books that explores the complex and ever-changing relationship between tourism and culture(s). The series focuses on the ways that places, peoples, pasts and ways of life are increasingly shaped/transformed/created/packaged for touristic purposes. The series examines the ways tourism utilises/makes and re-makes cultural capital in its various guises (visual and performing arts, crafts, festivals, built heritage, cuisine, etc.) and the multifarious political, economic, social and ethical issues that are raised as a consequence.

Understanding tourism's relationships with culture(s) and vice versa, is of ever-increasing significance in a globalising world. This series will critically examine the dynamic inter-relationships between tourism and culture(s). Theoretical explorations, research-informed analyses and detailed historical reviews from a variety of disciplinary perspectives are invited to consider such relationships.

Full details of all the books in this series and of all our other publications can be found on http://www.channelviewpublications.com, or by writing to Channel View Publications, St Nicholas House, 31–34 High Street, Bristol BS1 2AW, UK.

TOURISM AND CULTURAL CHANGE: 47

Cultural Tourism in Southern Africa

Edited by
Haretsebe Manwa, Naomi Moswete and Jarkko Saarinen

CHANNEL VIEW PUBLICATIONS
Bristol • Buffalo • Toronto

Library of Congress Cataloging in Publication Data
Names: Manwa, Haretsebe, editor.|Moswete, Naomi, editor.|
Saarinen, Jarkko, 1968- editor.
Title: Cultural Tourism in Southern Africa/Edited by Haretsebe Manwa, Naomi
Moswete and Jarkko Saarinen.
Description: Buffalo, NY : Channel View Publications, 2016. |
Series: Tourism and Cultural Change: 47 | Includes bibliographical references and index.
Identifiers: LCCN 2015029098| ISBN 9781845415525 (hbk : alk. paper) | ISBN
 9781845415518 (pbk : alk. paper) | ISBN 9781845415532 (ebook)
Subjects: LCSH: Heritage tourism—Africa, Southern.
Classification: LCC G155.A356 C85 2016 | DDC 338.4/79168—dc23 LC record available
at http://lccn.loc.gov/2015029098

British Library Cataloguing in Publication Data
A catalogue entry for this book is available from the British Library.

ISBN-13: 978-1-84541-552-5 (hbk)
ISBN-13: 978-1-84541-551-8 (pbk)

Channel View Publications
UK: St Nicholas House, 31–34 High Street, Bristol BS1 2AW, UK.
USA: UTP, 2250 Military Road, Tonawanda, NY 14150, USA.
Canada: UTP, 5201 Dufferin Street, North York, Ontario M3H 5T8, Canada.

Website: www.channelviewpublications.com
Twitter: Channel_View
Facebook: https://www.facebook.com/channelviewpublications
Blog: www.channelviewpublications.wordpress.com

Copyright © 2016 Haretsebe Manwa, Naomi Moswete, Jarkko Saarinen and the authors
of individual chapters.

All rights reserved. No part of this work may be reproduced in any form or by any means
without permission in writing from the publisher.

Typeset by Techset Composition India(P) Ltd, Bangalore and Chennai, India.
Printed and bound in Great Britain by Short Run Press Ltd.

Contents

Cases and Issues		vii
Figures, Tables and Plates		ix
Abbreviations		xi
Contributors		xv
Preface		xix

Part 1: Perspectives on Cultural Tourism

1 Introduction   3
  Haretsebe Manwa, Naomi Moswete and Jarkko Saarinen

2 Cultural Tourism in Southern Africa: The Role of Local Cultures and Ethnicity in Tourism Development   17
  Jarkko Saarinen

3 Integrating Indigenous Knowledge in the Development of Cultural Tourism in Lesotho   31
  Tsitso Monaheng

4 Narrative and Emotions: Interpreting Tourists' Experiences of Cultural Heritage Sites in KwaZulu-Natal   47
  Joram Ndlovu

5 Cultural Heritage Tourism Development in Post-Apartheid South Africa: Critical Issues and Challenges   58
  Gareth Butler and Milena Ivanovic

6 Cultural Tourism and the Arts Festivals   76
  Corné Pretorius

7   Reflections on International Carnivals as a Destination
        Recovery Strategy: The Case of Zimbabwe   86
    Cleophas Njerekai

Part 2: Impacts and Management of Cultural Tourism

8   The Commodification of World Heritage Sites:
        The Case Study of Tsodilo Hills in Botswana   101
    Joseph E. Mbaiwa

9   Tourism and the Social Construction of Otherness
        through Traditional Music and Dance in Zimbabwe   121
    Patrick Walter Mamimine and Enes Madzikatire

10 Rural Cultural Tourism Development and Agriculture:
        Evidence from Residents of Mmatshumu Village
        in the Boteti Region of Botswana   132
    Monkgogi Lenao

11 From Hunting-Gathering to Hospitality? Livelihoods
        and Tourism Use of Bushman Paintings in
        the Brandberg Mountain, Namibia   145
    Renaud Lapeyre

12 Emergence of Cultural Tourism in Southern Africa:
        Case Studies of Two Communities in Botswana   165
    Masego Monare, Naomi Moswete, Jeremy Perkins and Jarkko Saarinen

13 Cultural Tourism in Southern Africa: Progress,
        Opportunities and Challenges   181
    Naomi Moswete, Jarkko Saarinen and Haretsebe Manwa

    Index   190

Cases and Issues

Box 3.1	Cultural products: A driver for informal sector business tourism in southern Africa	34
Box 5.1	Industrial heritage tourism – the 'Big Hole', Kimberley, South Africa	61
Box 8.1	Authenticity of the Basotho hat	104
Box 11.1	Cultural village tourism in Namibia: The case of Helvi Mpingana Kondombolo Cultural Village	147
Box 12.1	Authenticity in tourism	166

Figures, Tables and Plates

Figures

Figure 6.1	Relationship between cultural tourism and the arts festival	79
Figure 7.1	Part of the cultural diversity that has characterized Zimbabwe's international carnivals since 2013	93
Figure 8.1	Map of Botswana showing the location of Tsodilo Hills and Okavango Delta	107
Figure 11.1	The Tsiseb conservancy and the Brandberg Mountain	146
Figure 12.1	Map of Botswana showing study site geographical location	173

Tables

Table 1.1	Doxey's irritation index (irridex) model	5
Table 5.1	Purpose of visit to South Africa, 2012 (international tourists)	60
Table 8.1	Stakeholders and stakeholder interest at Tsodilo Hills	110
Table 8.2	Projects sponsored by Diamond Trust at Tsodilo Hills	111
Table 8.3	Impacts and benefits to stakeholders	114
Table 8.4	Potential benefits at Tsodilo Hills	116
Table 8.5	Summary of environmental threats at Tsodilo Hills	117

Table 11.1	Working conditions as reported by employees interviewed	157
Table 11.2	Employees' previous professional occupations	160
Table 11.3	Lodge revenue as the main source of income for employees	160
Table 11.4	Employees' support for their relatives and dependents	161
Table 11.5	Employees' personal spending	162
Table 12.1	Examples of cultural heritage attractions in southern Botswana	170
Table 12.2	Cultural villages for tourism in Botswana	172

Plates

Plate 2.1	The Lesedi Cultural Village, South Africa	25
Plate 5.1	The Big Hole, 2013	63
Plate 8.1	The Basotho hat	105
Plate 11.1	The entrance to the Helvi Ya Mpingana Kondombolo Cultural Village	149
Plate 12.1	Craft market development at Hartbeespoort Dam, South Africa	168
Plate 12.2	Bahurutshe cultural village in Mmankgodi, Botswana. The village provides accommodation and cultural and touristic elements, including 'edutainment', integrating entertainment with cultural education aspects.	175

Abbreviations

ANC	African National Congress
ATLAS	Association for Tourism and Leisure Education
AWF	African Wildlife Fund
BTDP	Botswana Tourism Development Programme
CAMPFIRE	Communal Area Management Programme for Indigenous Resources
CBNRM	Community Based Natural Resources Management
CBOs	Community Based Organisation
CES	Centre de Estudes Socias
CHT	Cultural Heritage Tourism
CMT	Conventional Mass Tourism
DEAT	Department of Environmental Affairs and Tourism
DMNM	Department of Museum and National Monuments
DOT	Department of Tourism
DTI	Department of Trade and Industry
EDD	Economic Development Department
EDZ	Exclusive Development Zone
EIA	Environmental Impact Assessment
EMA	Environmental Management Authority
ENP	Etosha National Park
ESD	Education for Sustainable Development Programme
EU-NTDP	European Union Namibia Tourism Development Programme
FEE	Festivals, Events and Exhibitions Department
FIFA	The Fédération Internationale de Football Association
FGDs	Focus Group Discussions
GCT	Gaing'O Community Trust
GoB	Government of Botswana
HIC	Harare International Carnival

IFACCA	International Federation of Arts Councils and Culture Agency
IFTR	The International Federation for Theatre Research
IKS	Indigenous Knowledge Systems
ILO	International Labour Organization
IUCN	International Union for Conservation of Nature
JMC	Joint Management Committee
KZN	KwaZulu-Natal
LED	Local Economic Development
LHR	Liberation Heritage Route
MDTP	Maluti-Drakensberg Transfrontier Programme
MET	Ministry of Environment and Tourism
MICE	Meetings, Incentives, Conventions and Exhibitions
MYSC	Ministry of Youth, Sports and Culture
NACOBTA	Namibian Community-Based Tourism Association
NACSO	Namibian Association of Community Based Natural Resource Management Support Organisations
NDP	National Development Plan
NDT	National Department of Tourism (Republic of South Africa)
NGO	Non-Governmental Organisation
NTDP	Namibia Tourism Development Programme
NTSS	National Tourism Service Strategy (South Africa)
OECD	Organisation for Economic Cooperation and Development
RIM	Robben Island Museum (South Africa)
RISE	Rural Institute for Social Empowerment
SA	South Africa
SADC	Southern Africa Development Community
SAT	South African Tourism
SDI	Spatial Development Initiative
SET	Social Exchange Theory
SIT	Special Interest Tourism(s)
SMEs	Small and Medium-Sized Enterprises
TCT	Tsodilo Community Trust
TDF	Tourism Development Framework
TWHMP	Tsodilo World Heritage Management Plan
UN	United Nations
UNDP	United Nations Development Program
UNESCO	United Nations Scientific and Cultural Education Council
UNWTO	United Nations World Tourism Organisation
USAID	Unites States Agency for International Development

UWCC	Ugab Wilderness Community Campsite
VDC	Village Development Committee
WHS	World Heritage Site
WTO	World Tourism Organisation
WTTC	World Travel and Tourism Council
WWF	World Wildlife Fund for Nature
ZTA	Zimbabwe Tourism Authority
4Hs	Habitat, History, Handicrafts and Heritage

Contributors

Gareth Butler is a lecturer in International Tourism Management at Flinders University, Australia, and Senior Research Affiliate at the University of Johannesburg, South Africa. His research interests include tourism mobilities, tourism and sustainability, and cultural heritage tourism development.

Milena Ivanovic is a lecturer in Cultural Tourism and Tourism Development at the University of Johannesburg, South Africa. Her research interests include authenticity of the tourism experience, transmodernism and cultural heritage tourism development and management.

Mary-Ellen Kimaro is a lecturer in Tourism Management, Tourism Planning and Tourism Marketing at the University of Namibia. Her research interests include perception studies, heritage tourism, tourism planning and development and tourism management.

Renaud Lapeyre is the scientific coordinator of the INVALUABLE project and a Research Fellow at the International Institute for Sustainable Development and International Relations (IDDRI) in Paris, France. His doctoral dissertation analysed collective action and socio-economic impacts of community-based natural resource management (CBNRM) in Namibia. Currently he researches market-based instruments for biodiversity conservation.

Monkgogi Lenao is a lecturer in the Department of Tourism and Hospitality Management at the University of Botswana. Until December 2015 Lenao was a postdoctoral research fellow at the University of Oulu. Research interests include: culture and heritage tourism, community-based tourism, tourism and rural development and border studies in tourism.

Enes Madzikatire holds a Master of Philosophy degree in Hospitality and Tourism, a Bachelor of Technology degree in Education, a Technical and

Vocational Diploma in Education and is currently studying for a Doctor of Philosophy degree in Hospitality and Tourism with the Chinhoyi University of Technology in Zimbabwe. Enes is a national consultant for the cosmetology and hairdressing programmes in Zimbabwe. Research interests include: ethnicity and culture, spa technology, spa and wellness tourism, self-presentation and deportment and hospitality aesthetics.

Patrick Walter Mamimine holds a PhD in Anthropology and Sociology of Tourism from the University of Zimbabwe. He is Chairperson and teaches tourism courses in the Department of Hospitality and Tourism at the Chinhoyi University of Technology in Zimbabwe. He is also a regional tourism consultant and currently serves as the World Bank Tourism Consultant to the Government of Botswana on the Northern Botswana Human Wildlife Co-Existence Project. Research interests include: cultural tourism, community-based tourism, community-based natural resource management, transfrontier tourism planning and development, ecotourism, tourism and terrorism, township tourism and urban tourism.

Haretsebe Manwa is an associate professor and Head/Programme Leader of the Tourism Department at North West University, Mafikeng Campus, South Africa. Her research interests include wildlife tourism, community-based tourism, governance and tourism planning and development, destination development, cultural tourism, tourism and poverty alleviation and gender and tourism.

Joseph E. Mbaiwa is a professor of Tourism Studies at the Okavango Research Institute, University of Botswana. Professor Mbaiwa is also a Research Affiliate at the School of Tourism & Hospitality, Faculty of Management, University of Johannesburg, South Africa. His research interests are tourism development, conservation and rural livelihood development. He holds a PhD in Park, Recreation and Tourism Sciences.

Tsitso Monaheng is an associate professor in the Department of Development Studies at the University of the North-West, Mafikeng Campus, South Africa. He holds a Doctor of Literature and Philosophy Degree (DLitt et Phil) and has 15 years' experience as an academic. He also has practical experience in the field of community development, which is where his main academic interest lies.

Masego Monare is a social studies teacher at Moselewapula Junior School, Gaborone. Monare has completed a Master's in Environmental Science at the

University of Botswana. Monare's research interests include: cultural tourism, agricultural tourism and environmental education.

Naomi Moswete is a senior lecturer in the Department of Environmental Science, University of Botswana and also an interim coordinator for the International Tourism Research Centre at the University of Botswana. Moswete's research interests include tourism as a strategy for rural development, community-based natural resources management, community-based ecotourism; trans-frontier park-based tourism, community conservation, natural and heritage resource management and gender-based empowerment via tourism in Africa and other countries.

Joram Ndlovu holds a PhD in Tourism Management from the University of Pretoria, South Africa. He has vast experienced in teaching tourism at tertiary institutions. He is currently the Culture Cluster Leader in the School of Social Sciences at the University of KwaZulu-Natal. His research interests are food, culture and identity.

Hilma Joolokeni Nengola holds a BA in Tourism Management. She is currently working with the tourism industry in north-east Namibia.

Cleophas Njerekai is a PhD candidate at Midlands State University in Zimbabwe. He is currently a lecturer in tourism at the same university. He holds an MSc in Tourism from the University of Zimbabwe. His research interests include: cultural tourism, green tourism, tourism entrepreneurship.

Jeremy Perkins is an associate professor in range ecology with the Department of Environmental Science, University of Botswana. His research interests are livestock and wildlife management on rangelands, biodiversity conservation, environmental flows in sand rivers, CBNRM and environmental impact assessment.

Corné Pretorius holds a PhD in Events Tourism from the North West University, Potchefstroom Campus, South Africa. She is a lecturer in the Department of Tourism, North West University, Mafikeng Campus. Her research interests are: art and cultural tourism, events tourism and tourism marketing.

Christian M. Rogerson is a professor in the School of Tourism & Hospitality, Faculty of Management, University of Johannesburg, South Africa. His major research interests are centred on the tourism–development nexus with particular concern for issues of local and regional economic development, small enterprise development and spatial change.

Jarkko Saarinen is Professor of Geography at the University of Oulu, Finland, and Distinguished Visitor Professor at the University of Johannesburg, South Africa. His research interests include tourism and development, sustainability and responsibility in tourism, tourism–community relations, tourism and climate change, community-based natural resource management and wilderness studies.

Clinton David van der Merwe is a lecturer of geography, at the School of Education, University of the Witwatersrand, Johannesburg, South Africa, and a PhD candidate, reading heritage tourism, in the Department of Geography, Environmental Management and Energy Studies, University of Johannesburg. His research interests are: urban renewal, environmental justice, and geographical education.

Preface

Cultural tourism has generated increasing research interest in the context of southern Africa but there have not been any major academic attempts to create an overview of, or to discuss and critically evaluate the field more widely and beyond South Africa.

In general, the role of international and domestic tourism has grown in southern Africa and tourism has become a major industry in the region. Tourism is also becoming an important field of study and research at the region's universities. Southern Africa poses an interesting context and target for many international scholars and universities, and a number of international universities have located satellite campuses and/or programmes in the region (e.g. Monash University, Australia, and the University of Birmingham, UK). There are also various collaborative/international links between southern African universities and international universities in tourism research and teaching which demonstrates the importance of and growing interest in the region's tourism base.

In addition to wildlife and natural attractions, the southern Africa region has diverse and numerous ethnic groups, languages, communities, traditions, heritage, religions, museums, townships, battlefields, San paintings, rural landscapes, cuisine, vineyards, etc. This makes the region an ideal destination in which to develop and market cultural and heritage tourism. As a result of the increasing roles of culture and people, many countries in the region regard the promotion of tourism as a viable strategy that can be used to attract visitors, capital and foreign investment to the country by utilising both wildlife and local cultures for tourism. There is also 'a strong focus on the potential of cultural tourism to contribute towards the goals of sustainable rural development'.

All this calls for more research-based higher education materials on the nature, role, possibilities and challenges of the utilisation of culture and local communities in tourism. Many of the region's cultural attractions are sensitive to the changes caused by uncontrolled tourism development, which demands increasing knowledge on the use, impacts and management of cultural tourism. There is no comprehensive academic account of cultural tourism in southern Africa that is easily accessible to higher education students and scholars. This book provides an accessible introduction to and discussions on southern African cultural tourism issues complete with case studies.

The aim of this book therefore is to introduce the basic elements of cultural tourism by placing both theoretically and empirically based discussions in the southern African context. The purpose being to provide deep and wide-ranging perspectives on the utilisation of culture in southern African tourism and on the related impacts, possibilities and challenges. This book will serve higher undergraduate (3/4) level and postgraduate level students in tourism studies, tourism management, human and cultural geography, sociology/anthropology, and development studies.

A number of individuals and institutions have contributed in different ways to this book project. We would like to acknowledge the home universities of the authors who allowed the contributors time off to embark on this project. We would like to extend a special thank you to Sarah Williams and the Channel View staff and Mike Robinson and Alison Phipps (series editors) for their interest, support and great help with the book, starting from the early stages of the project till the printing of the finished product in hand. We would also like to thank the anonymous evaluators and reviewers of the book proposal and the outcome. Last, but not least, we would like to thank our family members for their support.

Haretsebe Manwa, Naomi Moswete and Jarkko Saarinen

Part 1

Perspectives on Cultural Tourism

1 Introduction

Haretsebe Manwa, Naomi Moswete and Jarkko Saarinen

Introduction

Traditionally the southern African region's tourist product has been strongly dependent on the natural environment and wildlife. However, southern Africa is endowed with a wealth of, and diverse cultural resources that include, but are not limited to, ethnic groups, languages, communities, traditions, heritage, religions, museums, townships, battlefields, San paintings, rural landscapes, cuisine, vineyards, etc. Indeed, the attraction and recent success of southern Africa in tourism development has been based on its diversity rather than on a single 'product'. The White Paper on Development and Promotion of Tourism in South Africa (DEAT, 1996), for example, states that the attractiveness of the region is based on relatively accessible wildlife, beautiful scenery, unspoiled nature and diverse traditional and township cultures (see also Republic of Namibia, 1994; Robinson, 2001; South Africa National Heritage and Cultural Tourism Strategy, 2012). The White Paper suggests that the competitive advantage of the country is no longer based on natural elements only, but increasingly includes man-made environments. This notion refers to the increasing role of cultural tourism and local communities in the future of tourism in the region. The emphasis on cultural tourism has been a central objective of post-apartheid tourism policy in South Africa since 1994 (see Van Veuren, 2001). This had an influence on the wider tourism policy of the Southern Africa Development Community (SADC) and nowadays tourism is viewed as an essential sector of regional and national reconstruction and development in the region (see Binns & Nel, 2002; Rogerson, 1997; Visser & Rogerson, 2004). The Spatial Development Initiative (SDI) in South Africa, for example, is aimed at economic empowerment of local communities and places strong emphasis on cultural tourism.

As a result of the increasing roles of culture and people, many countries in the region regard the promotion of tourism as a viable strategy that can be used to attract visitors, capital and foreign investment by utilising both wildlife and local cultures for tourism (see UNWTO, 2008). This diversification trend towards people and local cultures, however, is relatively new in the region and the utilisation of culture and community-based tourism is often characterised as being complementary to wildlife, safari and wilderness tourism (Manwa, 2007). However, the growing importance of culture and people as tourist attractions is evident already. For example, over one-third of international tourists visit a cultural village during their stay in South Africa (Department of Tourism, 2014; van Veuren, 2004: 140).

Conceptualising Cultural Tourism

Although cultural tourism has interested the attention of researchers for a long time, there has been no agreement regarding what cultural tourism entails (Debeş, 2011; Hausmann, 2007; Kastenholz *et al.*, 2013; Smith, 2009). This is also reflected in the subjects covered in the chapters of this book, which range from arts festivals to carnivals. Hausmann (2007) outlines some of the commonalities in most definitions of cultural tourism:

(1) Cultural tourism is a form of special interest. It is contended that cultural tourists do not differ from other tourists in that they also have limited time and resources.
(2) Cultural tourism refers to the use of heritage sites and their offerings and value to visitors.
(3) Like other tourists, cultural tourists wish to consume a variety of culture-related services and experiences.
(4) Cultural tourism must consider the preferences of the tourists.
(5) Cultural tourism is a complex phenomenon.

Richards (1996: 24) has defined cultural tourism as 'all movement of persons to specific cultural attractions, such as heritage sites, artistic and cultural manifestations, arts and drama outside their normal place of residence'. Thus, the tangible and intangible cultural elements of destination areas are the key resources for cultural tourism and the cultural tourists' motivations.

Cultural tourism and the global-local nexus

Due to the increasing attractiveness and promotion of culture and people, many rural places, communities and cultures in the region are currently tied to global economies and larger cultural and political networks through the development of tourism and related socio-economic activities. Travelling for cultural experience has always been part and parcel of tourism, starting with the Grand Tour (Hibbert, 1969). The Grand Tour was the manifestation of the globalisation of cultural tourism; often associated with the elite, young men from the British aristocratic classes and from other parts of northern Europe travelled to continental Europe for educational and cultural purposes (Weaver & Lawton, 2006: 61). Undertaking such a tour enhances the participant's status as a member of the elite (Burkhart & Medlik, 1981: 4).

Increasingly, Western society is alienating its people, a sense of self is being lost. As a consequence, people are seeking escape from the stressful everyday in other places and cultures. Travel to 'exotic' places which have not been affected by modernity provides novelty and existential authenticity, by being true to oneself in strange cultures and with strange people (Boorstin, 1969; Cohen, 1979; MacCannell, 1973; Uriel, 2005; Wang, 1999). People from more culturally distant places are motivated to travel for cultural reasons and seek deeper experiences, whereas tourists from culturally proximate regions are less interested in cultural tourism and seek superficial entertainment-oriented experiences (McKercher & du Cros, 2003: 46).

The interaction between hosts and guests can result in a number of impacts (Wall & Mathieson, 2006). Some scholars, such as Doxey (1976), see the interaction as a continuum of the level of local community's tolerance of tourists (Table 1.1). According to Doxey, the early stages of contact are characterised by indifference among residents regarding the presence of tourists, which grows to irritation and hostility as the tourism destination develops into a mass tourist destination.

There have been varied and often contrasting reactions to Doxey's theory. For example, Mathieson and Wall (1982) note that the volume of

Table 1.1 Doxey's irritation index (irridex) model

Host perceptions of tourists	Characterisation
Euphoria	Visitors are welcome and perceived very positively
Apathy	Visitors are taken for granted. Host–guest relations become formalised
Annoyance	Saturation point where the local people have doubts about evolving tourism
Antagonism	Open expression of irritation, even hostility, that influences the reputation of the destination negatively

Source: Doxey (1976).

tourists in comparison to the local population impacts on attitudes towards tourists. In cases where the number of tourists is greater than the number of local residents, there is bound to be hostility towards tourists because the local community is overwhelmed access to basic amenities becomes restricted (Horn & Simmons, 2002). Racial composition of tourists, cultural background and socio-cultural differences between tourists and residents are other factors which can influence cultural understanding between hosts and guests (Faulkner & Tideswell, 1997; Huimin & Ryan, 2012; Ryan et al., 2011).

Other views discount the theory that transformation of a destination into a mass tourism destination negatively influences the hosts' reaction to tourists. Instead, they attribute negative attitudes to the lack of benefits from tourism development. This is the basis of the Social Exchange Theory (SET) (McGehee & Andreck, 2004). The proponents of SET argue that communities will support tourism only if the benefits derived from it exceed the costs that the community has to endure. Tangible benefits such as employment, empowerment or direct participation in the tourism industry are a definite guarantee of support for tourism development (Nunkoo & Gursoy, 2012; Nunkoo & Ramkissoon, 2011, 2012; Ward & Berno, 2011).

People generally have preconceived ideas about foreign cultures which have culminated in prejudices and fears. Cultural interaction between hosts and guests can result in cultural understanding which fosters peace and global understanding among people from different parts of the world (Besculides et al., 2002; Borowiecki & Castiglione, 2014; Nyaupane et al., 2008; Pearce, 1995; Simpson, 2008; Suntikul et al., 2010). Once there is interaction and understanding of other cultures a feeling of appreciation and understanding of self, one's country and other cultures ensues (Cohen, 1979; Coulson et al., 2014; McKercher & Chow, 2001; Pearce, 1995, 2010; Yu & Lee, 2014: 235).

Other scholars contest the view of cultural understanding, instead they contend that interaction can re-enforce held prejudices towards other cultures (Yang, 2011). An important point is raised which is pertinent in developing economies, especially in sub-Saharan Africa, where most rural communities are barely literate and cannot meaningfully interact with tourists. Those representing the local people and their culture often portray a staged and inauthentic culture of the local people. Salazar's (2012) studies in Tanzania point to the lack of representation and interpretation of the local Maasai culture as an example of this. The tour guides and communities that were presenting the Maasai culture to tourists belonged to the Meru tribe. They staged a fake Maasai culture, banking on the widely held myths and prejudices about the Maasai. In the process they denigrated the Maasai and depicted them as backward. In this instance there was no cultural exchange between hosts and guests.

This is in contrast to experiences of indigenous peoples in developed economies like Australia and Canada. In these countries, aboriginal people have successfully used cultural tourism to valorise cultural identity and heritage (Debeş, 2011). Cultural tourism has empowered indigenous communities as they use it to effectively educate tourists about the aboriginal culture. Moscardo *et al.* (2013: 552) caution against assuming that in developed countries like Australia interaction with tourists always brings positive exchanges. Their studies in Queensland, Australia, have pointed to hostilities and resentment towards tourists, where residents have reported 'tourists being poor drivers, stories of culturally inappropriate or insensitive behaviours and the introduction of drug use and sexually transmitted diseases to residents'. Perhaps for cultural understanding and learning to take place there must not be too vast a difference between guests and host cultures (Jarvenpa, 1994).

Studies in different parts of the world have also shown that development of cultural tourism, especially in marginal areas, can be instrumental in creating jobs in areas with limited options for residents' productive employment (Besculides *et al.*, 2002). Residents also see tourism as a means of helping them to learn about, share and preserve their cultures (Besculides *et al.*, 2002; Butler *et al.*, 2014). For example, Butler *et al.*'s (2014: 199) studies in Malaysia note the pride of Malaysians from different ethnic backgrounds in religious-centred heritage attractions which have the potential to foster a collective national identity if promoted effectively.

Proponents of the development of cultural tourism justify it on the basis of bringing about sustainable rural development (van Veuren, 2004: 139; Salazar, 2012). Some tangible developments include better road infrastructure, better housing and consequent improvements in the standard of living of rural communities (Mbaiwa, 2005; Zamani-Farahani & Musa, 2012). In addition, it can revive traditional arts and crafts, traditional festivals, songs, music and dance, restore historical sites and monuments and demonstrate the effects and building of social capital (Mbaiwa, 2005; Mbaiwa & Sakuze, 2009; Mbaiwa & Stronza, 2010). More importantly, cultural tourists are less sensitive to price (Nicolau, 2010).

Cultural tourism is made up of diverse products which cater for varied and different market segments. There are products which attract high spenders from the middle classes, whom Kim *et al.* (2007) refer to as the 'highbrow culture' because of their preference for specific cultural products like art galleries, opera and concerts of classical music. Cultural tourism also attracts consumers from the working classes who consume 'lowbrow culture', such as movies, rock concerts and arts festivals (Kim *et al.*, 2007: 1370 citing Richards, 1996). The southern African region has diverse cultural products, characterised by unique cultures. For example, it is home to over 37 UNESCO

World Heritage Sites. South Africa alone has 11 official languages, representing its diverse cultures. Battlefields characterise a number of the southern African countries. These sites can be developed into 'dark tourism attractions'. A number of these countries also celebrate annual arts festivals, depicting different tribes with unique dance, song, music and food.

Cultural tourism does not only bring about positive impacts, it has also impacted communities negatively. Whilst the positive impacts must be promoted, negative impacts should also be recognised and mitigated in the development of cultural tourism in the southern African region. Some negative impacts have already been felt in some countries. Cultural tourism can result in the straining of amenities, thereby disadvantaging the locals in accessing those amenities. Instances like these happen when destinations exceed their carrying capacity. For example, in India, Tillotson (1988: 1940) observes that the over use of popular, prominent and accessible buildings results in their wear and tear.

Cultural tourism can result in a demonstration effect, where local communities, especially the youth, abandon their culture and start to emulate foreign cultures in the way they dress and the food they eat, etc. (Mbaiwa, 2011). Mbaiwa further notes that in the Okavango Delta, where community-based tourism has been successful, local communities have abandoned their traditional way of life; they now want, for example, to live in Western-style houses and eat Western food.

Demand for cultural products such as souvenirs, artefacts, crafts and traditional dances has resulted in commodification of cultural artefacts (Shepherd, 2002). Harrison (1994: 243–244) notes that given a monetary value, ritual and tradition become valueless for local inhabitants.

Another issue still challenging tourism development and cultural tourism in particular is the definition of a community. In most southern African countries there is a multiplicity of cultures, which makes it a challenge to decide whose culture should be included in the cultural product and therefore portrayed in advertising media. Second, is the issue of benefit sharing? Who should benefit from the cultural resources? Who interprets the culture? These issues touch on ethical considerations, power, empowerment, gender issues, participation and inequality. These are issues that will impact on sustainability of cultural tourism in the southern African region.

The chapters of this book use case studies to showcase some of the cultural tourism operational in southern African countries or types of products that have a potential for development. In Chapter 2, Jarkko Saarinen discusses the characteristic of cultural tourism in southern Africa by focusing on the role of local cultures and ethnicity in tourism development. His chapter conceptualises the specific fields of cultural tourism called ethnic and indigenous tourism, which are highly visible forms of cultural tourism in southern Africa. They are widely used in tourism promotion and products and also increasingly used by southern African governments to facilitate development, especially in rural areas and other peripheral parts. Saarinen notes that while such cultural tourism involves a 'promise' of development to localities, those ethnic and indigenous groups that are introduced to tourism are also fragile, especially when utilised in internationally oriented tourism operations. Therefore, the chapter concludes that tourism can provide development for local communities, but in order to do so, there is a serious need to include and implement a sustainable tourism development approach when integrating tourism and local ethnic and indigenous communities.

In Chapter 3, Tsitso Monaheng focuses on Lesotho. He presents a symbiotic relationship between culture and indigenous knowledge. Lesotho is rich in cultural resources. The sustainability of these resources is dependent on cultural tourism being driven by the Basotho people who are the custodians of the culture. Basotho people have used their indigenous knowledge to preserve their cultural resources. The chapter also emphasises the important role women play as custodians of both culture and indigenous knowledge systems. He advocates for women taking the lead in the development and marketing of cultural tourism in Lesotho.

Joram Ndlovu's chapter (Chapter 4) adopts Urry's (1990) tourist gaze to analyse tourists' motivation to visit cultural and heritage tourism sites in KwaZulu-Natal province, South Africa. Learning about self and others, satisfying curiosity, nostalgia, being part of the re-enactment of historical battlefields, and the way of life of the Zulu people are some of the motivations for visiting the province. He further emphasises accessibility of the cultural and heritage resources to both tourists and local communities. This is important for cultural exchange to take place. He notes, however, that cultural tourism is underplayed in destination marketing and there is a need to develop social capital so communities take leadership in developing and managing cultural and heritage tourism.

In Chapter 5, Gareth Butler and Milena Ivanovic approach the challenges that relate to the use of cultural heritage in tourism development in a problematic national identities context. Their focus is cultural heritage tourism development in post-apartheid South Africa. They provide an overview of current principles and practical challenges arising from the development of cultural heritage tourism in the country. While there are numerous strategies and policy documents supporting the utilisation of culture and heritage resources in tourism, the implementation and realities of cultural heritage tourism development are problematic. Butler and Ivanovic perceive that in an attempt to realign the nation's identity to be more reflective of Afrocentric discourse, successive South African governments have failed to develop a solution that is reflective of its highly diverse and transforming cultural heritage. They conclude that although there are painful memories of the apartheid era, which obviously need to be recognised, there is also a need to foster collective identities in domestic discourses that create feelings of 'togetherness' in cultural heritage tourism attractions and promote the country's rich cultural diversity.

Corné Pretorius, in Chapter 6, focuses on arts festivals as a form of cultural tourism in South Africa. South Africa has a lengthy experience of arts festivals being staged in different provinces. For example, wine and cheese festivals in the Eastern Cape, Nelson Mandela Bay Water Festival in the Eastern Cape, the Klein-Karoo National Arts Festival, the Aardklop National Festival and the Oppikoppi Music Festival. The chapter underlines the importance of the arts festival as a means of self-expression, national identity, pride in one's culture and cultural reengineering. Arts festivals can also lead to commodification of those aspects of culture which are popular with tourists and the discarding of less popular art forms.

In Chapter 7, Cleophas Njerekai discusses the International Harare Carnival which was introduced in Zimbabwe in 2013. Zimbabwe is battling with a battered international image as a tourist destination and is regarded as a pariah state. The introduction of the carnival will not only help revive the tourism industry but it is also a way of creating a positive image of the country through cultural exchange and interaction. Njerekai acknowledges that it is too early to measure the impacts of the carnival, which is in its second year. He, however, recognises that already both negative and positive impacts have been noted. Positive impacts include cultural exchange, national pride and development of small and medium-sized enterprises and the building of social capital. Negative impacts include limited buy-in by the local community, littering and traffic jams.

Joseph E. Mbaiwa, in Chapter 8, deals with cultural resource commodification, which has become a challenge for cultural resource managers and cultural tourism issues globally. He uses the concept of commodification to analyse the effects of cultural and/or tourism development on natural and cultural resources of a renowned World Heritage Site – Tsodilo Hills in Botswana. Further, he teases out how the introduction and/or growth of tourism has impacted on local livelihoods and conservation of the cultural-heritage resources of Tsodilo Hills. Some of the major findings at Tsodilo Hills are due to increased volumes of visitors, uncontrolled large groups and behaviours (school tours). In addition, socio-economic and environmental impacts uncovered include, but are not limited to, the overuse of resources by large religious groups (that is, water in the various 'springs', depletion of which is problematic); and uncontrolled behaviours and excavations, leading to vandalism at the site. Off-road driving by researchers, tourists, government workers and other visitors, as well as uncontrolled camping have also been noted as problems. Socio-economic impacts include unlimited accommodation facilities which have created competition for the locally operated campsite which forms a larger part of Tsodilo residents' livelihood. It is found that when tourism development occurs rapidly and without planning or appropriate regulatory control, it has the potential to make any heritage site, regardless of scale, vulnerable to over-utilisation and environmental degradation.

Patrick Walter Mamimine and Enes Madzikatire provide a critical narrative of the social construction of Otherness through the Shona people's traditional music and dance performed for a tourist audience at Chapungu Sculpture Park in Harare Zimbabwe in Chapter 9. They present the important role of the leader of the Chapungu Dance Group as a gatekeeper controlling the interaction process between the dance troupe and the audience in order to suit the expectations of the tourists. The chapter shows that the construction of images of authentic Otherness to a large extent succeeded because tourists were part of their construction through the positive feedback they gave of the projected images. The social construction of Otherness succeeded because of the leader's ability to constantly radiate images of a people whose culture is different from that of the tourists. The host continually demonstrated how the Shona culture was dissimilar to that of the tourists. To successfully construct Otherness the group leader needed a good knowledge of the culture of the majority of the visitors. He was therefore able to exploit his knowledge base by radiating images of cultural differences both historically and virtually.

In Chapter 10, Monkgogi Lenao analyses rural cultural tourism development using the case of the remote agriculture community of Mmatshumu village in the Boteti region of Botswana. In this chapter rural cultural tourism is discussed through the context of a community-based tourism approach. The chapter highlights the importance of community-based natural resources management (CBNRM) – a framework which has been adopted in Botswana and the rest of the southern African region to guide community rural tourism. In the study, community rural cultural tourism development and agriculture was explored, and the findings demonstrate positive and negative relationships between agriculture and rural cultural tourism development at Lekhubu Island, where cultural tourism aims to provide opportunities for local residents (jobs, markets for crafts and agricultural goods). There is a growing concern about the volume of livestock around the island, particularly near the campsites. Sightings of domestic animal and cattle droppings have an adverse impact on the cultural landscape of Lekhubu Island and reduce the quality of the visitor experience and visitor satisfaction. Such findings fit within the 'irridex' (*irritation*) model that presents the social impacts of tourism with reference to tourism being introduced into rural communities. The chapter concludes by stating the potential for both positive and negative coexistence between agriculture and tourism development at Lekhubu Island. This coexistence could be reinforced by educating local people about the importance of the conservation and preservation of Lekhubu Island for posterity and to safeguard its unique cultural landscape. In addition it could encourage livestock rearing in nearby settlements because pastoral farming plays a vital role in the history and cultural landscape of the area.

In Chapter 11 Renaud Lapeyre analyses the touristic uses of 'bushman paintings' in Brandberg Mountain, Namibia. He focuses on tourism accommodation activities that are dependent on cultural tourism resources in the Tsiseb Conservancy with the specific aims of assessing the impact of cultural tourism on local community livelihoods and the level of empowerment involved with tourism in the conservancy context. Lapeyre utilises Ostrom's (1990) institutional analysis with a sustainable livelihood approach. He concludes that there is a relatively limited level of local participation and involvement in cultural tourism development and related decisions in the area. The local community is a passive landowner that is basically renting its land and natural and cultural resources without any further agency in tourism operations and development. However, Lapeyre also concludes that beyond empowerment and participation issues, which are crucial for the sustainability of tourism development in a long-term perspective, the evolving tourism industry has provided possibilities for local households to improve their livelihoods in the region where very limited alternative options exist.

Monare *et al.* (Chapter 12) draw readers' attention to the emergence of cultural tourism in southern Africa with a focus on Botswana. In this chapter the authors showcase the development of cultural tourism 'for tourism' in remote parts of most southern African countries. They discuss the emergence of, and compare, cultural tourism development in two communities (the Mmankgodi and Kanye villages) in southern Botswana. The chapter interrogates the following questions: How and why were the cultural villages created? Who owns the villages? Do both villages offer cultural tourism in the context of Botswana or as understood in the countryside? What type of activities and services are offered? And, who are the key stakeholders? The findings include that there are similarities and differences between the two cultural villages. In addition, cultural tourism is new to Botswana, yet a few private and community-owned cultural villages for tourism exist. These villages play a key role in safeguarding the Balete and Bangwaketse cultures and Batswana heritage, as well as providing employment opportunities and/or forming recreation and tourism options for local people and the international community. The chapter closes by revealing that the two cultural villages are private and family-owned projects, and that there is low involvement and minimal benefits for the residents of both Mmankgodi and Kanye. In this light the authors recommend the introduction of community-based cultural tourism that would be communally managed and provide equitable benefits.

References

Besculides, A., Lee, M.E. and McCormick, P.J. (2002) Residents' perceptions of the cultural benefits of tourism. *Annals of Tourism Research* 29 (2), 303–319.
Binns, T. and Nel, E. (2002) Tourism as a local development strategy in South Africa. *The Geographical Journal* 168 (3), 235–247.
Boorstin, D.J. (1964) *The Image: A Guide to Pseudo-Events in America*. New York: Harper & Row.
Borowiecki, K.J. and Castiglione, C. (2014) Cultural participation and tourism flows: An empirical investigation of Italian provinces. *Tourism Economics* 20 (2), 241–262.
Burkhart, A. and Medlik, S. (1981) *Tourism: Past, Present and Future*. London: Butterworth-Heinemann.
Butler, G., Khoo-Lattimore, C. and Mura, P. (2014) Heritage tourism in Malaysia: Fostering a collective national identity in an ethnically diverse country. *Asia Pacific Journal of Tourism Research* 19 (2), 199–218. DOI: 10.1080/10941665.2012.735682.
Cohen, E. (1979) Phenomenology of tourist experiences. *Sociology* 13 (2), 179–201.
Coulson, A.B., MacLaren, A.C., McKenzie, S. and O'Gorman, K.D. (2014) Hospitality codes and Social Exchange Theory: The Pashtunwali and tourism in Afghanistan. *Tourism Management* 45, 134–141.
DEAT (Department of Environmental Affairs and Tourism) (1996) *White Paper: The Development and Promotion of Tourism in South Africa*. Pretoria: Government of South Africa.
Debeş, T. (2011) Cultural tourism: A neglected dimension of tourism industry. *Anatolia: An International Journal of Tourism and Hospitality Research* 22 (2), 234–251. DOI: 10.1080/13032917.2011.593910.
Department of Tourism (2014) Annual Report 2013/2014. See www.tourism.gov.za/AboutNDT/Publications/NDT%20Annual%20Report%202013_14.pdf (accessed 27 January 2015).
Doxey, G. (1976) When enough's enough: The natives and residents in old Niagra. *Heritage Canada* 2 (2), 26–27.
Faulkner, B. and Tideswell, C. (1997) A framework for monitoring community impacts of tourism. *Journal of Sustainable Tourism* 5 (1), 3–28.
Harrison, D. (1994) Learning from the Old South by the New South? The case of tourism. *Third World Quarterly* 15 (4), 707–721.
Hausmann, A. (2007) Cultural tourism: Marketing challenges and opportunities for German cultural heritage. *International Journal of Heritage Studies* 13 (2), 170–184.
Hibbert, C. (1969) *The Grand Tour*. London: Weidenfeld and Nicholson.
Horn, C. and Simmons, D. (2002) Community adaptation to tourism: Comparison between Rotorua and Kaikoura, New Zealand. *Tourism Management* 23, 133–143.

Huimin, G. and Ryan, C. (2012) Tourism destination evolution: A comparative study of Shi Cha Hai Beijing Hutong businesses' and residents' attitudes. *Journal of Sustainable Tourism* 20 (1), 23–40.

Jarvenpa, R. (1994) Commoditization versus cultural integration: Tourism and image building in the Klondike. *Arctic Anthropology* 31 (1), 26–46.

Kastenholz, E., Eusebio, C. and Carneiro, M.J. (2013) Studying factors influencing repeat visitation of cultural tourists. *Journal of Vocational Marketing* 19, 343–358.

Kim, H., Cheng, C.K. and O'Leary, J.T. (2007) Understanding participation patterns and trends in tourism cultural attractions. *Tourism Management* 28, 1366–1371.

MacCannell, D. (1973) Staged authenticity: Arrangements of social space in tourism settings. *The American Journal of Sociology* 79 (3), 589–603.

Manwa, H. (2007) Is Zimbabwe ready to venture into the cultural tourism market? *Development Southern Africa* 24 (3), 465–474.

Mathieson, A. and Wall, G. (1982) *Tourism, Economic, Physical and Social Impacts*. London: Longman.

Mbaiwa, J. (2005) The socio-cultural impacts of tourism in the Okavango Delta, Botswana. *Journal of Tourism and Cultural Change* 2 (3), 163–185.

Mbaiwa, J.E. (2011) Changes on traditional livelihood activities and lifestyles caused by tourism development in the Okavango Delta, Botswana. *Tourism Management* 32, 1050–1060.

Mbaiwa, J.E. and Sakuze, L.K. (2009) Cultural tourism and livelihood diversification: The case of Gcwihaba Caves and XaiXai village in the Okavango Delta, Botswana. *Journal of Tourism and Cultural Change* 7 (1), 61–75.

Mbaiwa, J.E. and Stronza, A.L. (2011) Changes in resident attitudes towards tourism development and conservation in the Okavango Delta, Botswana. *Journal of Environmental Management* 92, 1950–1959.

McGehee, N.G. and Andreck, K.L. (2004) Factors predicting rural residents' support for tourism. *Journal of Travel Research* 43, 131–134.

McKercher, B. and Chow, S.B. (2001) Cultural distance and participation in cultural tourism. *Pacific Tourism Review* 5, 21–30.

McKercher, B. and du Cros, H. (2003) Testing a cultural tourism typology. *International Journal of Tourism Research* 5, 45–58.

Moscardo, G., Konovalov, E., Murphy, L. and McGehee, N. (2013) Mobilities, community well-being and sustainable tourism. *Journal of Sustainable Tourism* 21 (4), 532–556.

Nicolau, J.L. (2010) Culture-sensitive tourists are more price insensitive. *Journal of Cultural Economics* 34 (3), 181–195.

Nunkoo, R. and Gursoy, D. (2012) Residents' support for tourism: An identity perspective. *Annals of Tourism Research* 39 (1), 243–268.

Nunkoo, R. and Ramkissoon, H. (2011) Residents' satisfaction with community attributes and support for tourism. *Journal of Hospitality and Tourism Research* 35 (2), 171–190.

Nunkoo, R. and Ramkissoon, H. (2012) Power, trust, social exchange and community support. *Annals of Tourism Research* 39 (2), 997–1023.

Nyaupane, G.P., Teye, V. and Paris, C. (2008) Innocents abroad. Attitude change toward hosts. *Annals of Tourism Research* 35 (3), 650–667.

Ostrom, E. (1990) *Governing the Commons. The Evolution of Institutions for Collective Action*. Cambridge: Cambridge University Press.

Pearce, P.L. (1995) From cultural shock and cultural arrogance to cultural exchange: Ideas towards sustainable socio-cultural tourism. *Journal of Sustainable Tourism* 3 (3), 143–154.

Pearce, P.L. (2010) New directions for considering tourists' attitudes towards Others. *Tourism Recreation Research* 35 (3), 251–258.
Republic of Namibia (1994) *White Paper on Tourism.* Windhoek: Cabinet of Namibia (29 March 1994).
Richards, G. (ed.) (1996) *Cultural Tourism in Europe.* Wallingford: CABI.
Robinson, J. (2001) Tourism development framework (TDF). Gaborone: United Nations Development Program and World Tourism Organization, Technical Missions to Botswana.
Rogerson, C.M. (1997) Local economic development and post-apartheid reconstruction in South Africa. *Singapore Journal of Tropical Geography* 18 (2), 175–195.
Ryan, C., Zhang, C. and Zeng, D. (2011) The impacts of tourism at a UNESCO heritage site in China a need for meta-narrative? The case of the Kaiping Dialou. *Journal of Sustainable Tourism* 19 (6), 747–765.
Salazar, N.B. (2012) Community-based cultural tourism: Issues, threats and opportunities. *Journal of Sustainable Tourism* 20 (1), 9–22.
Shepherd, R. (2002) Commodification, culture and tourism. *Tourist Studies* 2 (2), 183–201.
Simpson, M.C. (2008) Community benefit tourism initiative – A conceptual oxymoron? *Tourism Management* 29, 1–18.
Smith, M.K. (2009) *Issues in Cultural Tourism Studies* (2nd edn). London; New York: Routledge.
South Africa National Heritage and Cultural Tourism Strategy (2012) Department of Tourism, Republic of South Africa. See http://www.tourism.gov.za/AboutNDT/Branches1/domestic/Documents/National%20Heritage%20and%20Cultural%20Tourism%20Strategy.pdf (accessed 17 August 2015).
Suntikul, W., Bauer, T. and Song, H. (2010) Towards tourism: A Laotian perspective. *International Journal of Tourism Research* 12, 449–461.
Tillotson, S. (1988) Cultural tourism or cultural destruction? *Economic and Political Weekly* 23 (38), 1940–1941.
UNWTO (United Nations World Tourism Organisation) (2008) *Policy for the Growth and Development if Tourism in Botswana.* Gaborone: UNWTO and Department of Tourism, Botswana.
Uriely, N. (2005) The tourist experience: Conceptual development. *Annals of Tourism Research* 32 (1), 199–216.
Urry, J. (1990) *The Tourist Gaze.* London: Sage Publications.
Van Veuren, E.J. (2001) Transforming cultural villages in the spatial development initiatives of South Africa. *South African Geographical Journal* 83 (2), 137–148.
Van Veuren, E.J. (2004) Cultural village tourism in South Africa: Capitalising on indigenous culture. In C.M. Rogerson and G. Visser (eds) *Tourism and Development Issues in Contemporary South Africa* (pp. 139–160). Pretoria: Institute of South Africa.
Visser, V. and Rogerson, C.M. (2004) Researching the South African tourism and development nexus. *GeoJournal* 60, 201–215.
Wall, G. and Mathieson, A. (2006) *Tourism: Change, Impacts and Opportunities.* Harlow: Pearson Education.
Wang, N. (1999) Rethinking authenticity in tourism experience. *Annals of Tourism Research* 26 (2), 349–370.
Ward, C. and Berno, T. (2011) Beyond social exchange theory: Attitudes toward tourists. *Annals of Tourism Research* 38 (4), 1556–1569.
Weaver, D. and Lawton, L. (2006) *Tourism Management* (3rd edn). Melbourne: John Wiley & Sons.

Yang, L. (2011) Ethnic tourism and cultural representation. *Annals of Tourism Research* 38 (2), 561–585.
Yu, J. and Lee, T.J (2014) Impact of tourists' intercultural interactions. *Journal of Tourism Research* 52 (3), 225–238.
Zamani-Farahani, H. and Musa, G. (2012) The relationship between Islamic religiosity and residents' perceptions of socio-cultural impacts of tourism in Iran: Case studies of Sare'in and Masooleh. *Tourism Management* 33, 802–814.

2 Cultural Tourism in Southern Africa: The Role of Local Cultures and Ethnicity in Tourism Development

Jarkko Saarinen

Introduction

Cultural attractions have a long and important history in tourism, and currently they play an increasing role in tourism in many parts of the world (Richards, 1997; Smith, 2003). This makes it understandable why so many countries, places and communities use or aim to use culture and cultural tourism as a tool for socio-economic development (Smith & Richards, 2013). While the expansion of cultural tourism is based on the growth of the modern tourism industry, in general, it is also a product of (and a vehicle for) so-called cultural globalisation. The term cultural globalisation refers to an idea that globalisation is primarily about culture. In this context, Nijman (1999: 148) has defined cultural globalisation 'as acceleration in the exchange of cultural symbols among people around the world, to such an extent that it leads to changes in local popular cultures and identities'. Basically, this process is what cultural tourism is about: cultural exchange with tourists' motivation to experience different cultures and local needs to gain economically and/or in other ways from visiting tourists.

Conceptually, cultural tourism refers to the movement of people to cultural attractions away from their normal place of residence, with the intention to gather new information and experiences to satisfy their cultural needs (Richards, 1997: 25; Smith, 2003). Cultural tourism can include a wide variety of different activities, motivations, operations and attractions. In addition, cultural tourism can involve visiting tangible attractions and sites, such as heritage sites, museums, galleries, etc., but also experiencing intangible cultural components, such as traditions, performing arts, gastronomy and different kinds of lifestyles formed by values and religious activities of destination communities, for example (Smith & Richards, 2013).

Like many other forms of tourism, cultural tourism has a potential impact on the lives of hosting populations. In addition to being a tool for employment and revenue creation, cultural tourism can initiate cultural exchange and understanding, stimulate heritage conservation, and support local cultural sectors and identities in general (McKercher & du Cros, 2002; Richards, 1997). The development of cultural tourism can also assist indigenous people and ethnic minorities in showcasing their culture and revitalising their traditions (Grunewald, 2002; Mbaiwa & Sakuze, 2009; Santos & Yan, 2008). Obviously, if developed and promoted in unsustainable and irresponsible ways, cultural tourism can create negative consequences, such as commodification of local culture and traditions, demonstration effects (i.e. locals adopting tourists' cultural and consumption behaviours) and marginalisation of local cultures and communities (see Smith, 1989; Yang, 2011). Indeed, as indicated by Wu *et al.* (2014: 2) the aims 'to protect the values of ethnic culture and to generate economic gains' by utilising tourism 'may be incompatible' in many cases.

In recent years, many southern African countries and regional development actors have started to utilise local cultures, communities, traditions and heritage in tourism development (see Saarinen *et al.*, 2014). The development of cultural tourism and its various forms is seen as a good and often inexpensive way to attract tourism (see Rogerson & Visser, 2004), especially in peripheral areas (Rogerson, 2015). The southern African region offers a wide range of cultural tourism attractions and activities that are not limited 'only' to local traditional cultures but a 'fusion' of lifestyles and ways of living in urban settings and contemporary arts, for example (see Ivanovic, 2008). Indeed, most southern African countries have a rich, diverse and constantly transforming base for cultural offerings in tourism. Based on this, South African Tourism (SAT), for example, has identified the development of cultural tourism as one of the country's key growth areas (Ivanovic & Saayman, 2013). Similarly, Namibia's community-based tourism policy and Botswana's revised tourism policy place strong emphasis on community involvement and local culture(s) in tourism development (see Lapeyre, 2010; Saarinen *et al.*, 2014).

This chapter aims to focus on very specific fields of cultural tourism: ethnic and indigenous tourism. These forms of cultural tourism are highly visible in southern African tourism promotion and products, and as noted by Yang *et al.* (2008: 751) 'they are often used by governments to facilitate development'. However, while tourism involves a 'promise' of development of local communities, ethnic and indigenous groups are also fragile when utilised in tourism-related commerce (Smith, 1989; Wu *et al.*, 2014). The chapter starts by discussing the basic ideas and definitions of ethnic and indigenous tourism. After that some selected examples of ethnic tourism in southern Africa are discussed, followed by a concluding discussion emphasising the need to implement a sustainable tourism approach when aiming to integrate tourism and ethnic and/or indigenous communities.

Ethnic and Indigenous Tourism

Evolving role of ethnic and indigenous tourism

Many scholars have noted that there has been a growing interest in ethnic and indigenous tourism in the past two decades (see Hinch & Butler, 1996; Ryan, 2005). Cultural elements have attracted tourists for a long time, but relatively recent changes in the modes of tourism production and consumption have created substantial market conditions for new and varying forms of cultural tourism activities, such as ethnic and indigenous tourism (Saarinen, 2013). These changes in tourism are largely based on wider transformations in Western production from Fordism towards post-Fordism (Urry, 1990), i.e. from a mass scale to more individualised patterns of production and consumption. Within this context it is assumed that the new kind of tourism industry – 'new tourism' – can create more effective and flexible products that conform better to the new structures of individual preferences, motivations, segments and trends in tourism consumption (see Ryan *et al.*, 2000). Thus, this new tourism represents a new style of production (and consumption), with increasing flexibility, individuality and hybridity (see Poon, 1993), which has not only created better platforms for ethnic and indigenous tourism per se but also better possibilities to link them to various forms of tourism. In this respect a hybridisation of nature-based tourism, by adding 'complementing' cultural aspects to it, has a potential to benefit the southern African tourism industry and its development (see Manwa, 2003).

While the industry has transformed, ethnic communities and their living environments have also changed. In transforming rural economies many traditional livelihoods may have lost their ability to provide an economic basis for the well-being of the people, competing resource uses may have emerged, customary belief systems and knowledge may have eroded and the modernisation of the surrounding societies, in general, may have created external and internal pressures and conflicts among communities. At the same time, the ethnic and indigenous minorities have become increasingly aware of their role and position in societies and development (Ryan, 2005; Saarinen, 2013). These concerns and emerging awareness are supported by the development of international and national agreements on indigenous and other ethnic minority rights (see Greene, 2004; Sinclair, 2003).

The overall results of these changes in tourism and ethnic/indigenous people's environments are that:

(1) increasing and changing tourism demand is providing new possibilities for the development of tourism based on the attractiveness of ethnic and indigenous communities and cultures, and
(2) the institutional agreements and evolved critical awareness of the position and rights of cultural minorities have supported the bottom-up development in tourism and cultural conservation of ethnic and indigenous people.

These results or changes have positioned ethnic groups' agencies in new and often better ways when negotiating with tourism operations, their management and ownership questions. These changes are also highly intertwined; as noted by Wu *et al.* (2014: 5) 'economic gains and cultural conservation are not only core issues for indigenous tourism, but they are also inseparable'.

Integrated concepts

The definitions of ethnic and indigenous tourism are closely related. On the one hand Smith (1989: 2) has stated that ethnic tourism is the 'marketing of quaint customs of indigenous and often exotic peoples for tourist consumption'. On the other hand, from the indigenous tourism perspective, Smith (1996: 283) has indicated that indigenous tourism refers to tourism which 'directly involves native peoples whose ethnicity is a tourist attraction'. Based on this, ethnicity and indigeneity are highly intertwined concepts: ethnic and indigenous tourism can be understood as tourism activities in which ethnic groups and indigenous people are in a key role of attracting tourists and organising tourism activities. However, Yang (2011: 561) states that ethnic tourism 'refers to tourism motivated by a tourist's search for exotic cultural experiences, including visiting ethnic villages, minority homes and ethnic theme parks, being involved in ethnic events and festivals, watching traditional dances or ceremonies, or merely shopping for ethnic handicrafts and souvenirs'. Thus, ethnicity and ethnic tourism are not automatically linked to indigeneity and indigenous tourism. Indeed, while all indigenous tourism can be regarded as ethnic tourism, not all ethnic tourism is indigenous (i.e. not based on indigenous people but other ethnic groups and their role in tourism) (Yang & Wall, 2009).

The concept of indigenous people is a contested and highly debated issue (see Barnard, 2006; Kuper, 2003). In practice it can often involve ethnic minority groups (Saarinen, 2011, 2013; United Nations, 2007). Conceptually, ethnicity or an ethnic group refers to 'a socio-culturally distinct group of people who share a common history, culture, language, religion, and way of life' (Yang et al., 2008: 752), while indigenous people refers to a culturally distinct group of people that has been living in and occupying a region before the present group(s) dominated the region and prior to the creation of current nation states (Saugestad, 2001). This kind of definition of indigeneity relates to international agreements and policy documents developed by the United Nations (2004, 2007) and the International Labour Organization (ILO, 1989). The latter defines indigenous people as 'peoples in independent countries, who are regarded as indigenous on account of their descent from the populations which inhabited the country, or a geographical region to which the country belongs, at the time of conquest or colonisation or the establishment of present state boundaries and who, irrespective of their legal status, retain some or all of their own social, economic, cultural and political institutions' (ILO, 1989: 1; see also Barnard, 2006).

As there are potential conceptual differences between ethnic and indigenous tourism, this chapter utilises ethnic tourism to cover both ethnicity and indigeneity, while it uses indigenous tourism when the focus is specifically on tourism activities that are based on the attractiveness and role of indigenous people as defined above by the ILO (1989) and Saugestad (2001). In the context of sustainability, a key issue of ethnic and indigenous tourism relates how 'native' people are involved with tourism (Hinch & Butler, 2007; Saarinen, 2013). Smith (1996), for example, has identified four crucial elements in the development potential of ethnic/indigenous tourism: habitat, history, handicrafts and heritage (the 4Hs). These elements refer mainly to the cultural difference and attractiveness of native people, but they also indicate, at least indirectly, the needs related to ownership or control in indigenous tourism management. Indeed, many scholars include the element of control of indigenous but also of ethnic tourism. Hinch and Butler (2007: 5), for example, have strongly emphasised the importance of control as the key issue in indigenous tourism development. According to them indigenous tourism refers to operations in which indigenous people are directly involved, either through control and/or by having their culture serve as the essence of the attraction (see Li, 2006; Telfer & Sharpley, 2007). The element of control can be based on participatory planning, land and other resource leasing systems or (joint) ownership in business (Ashley & Roe, 2002; Saarinen, 2012), which aim to safeguard the outcome that ethnic groups would receive a significant share of the economic benefits of tourism in the forms of direct revenues and employment, upgraded infrastructures, the environment and housing standards, etc. (see Stronza, 2007).

In addition, as noted by Wu *et al.* (2014), the outcomes of cultural, ethnic and/or indigenous tourism are also dependent on the nature of visiting tourists. Related to this they have applied a concept of serious leisure to tourism by defining 'serious indigenous tourists'. These 'serious' tourists are visitors valuing indigenous cultures more highly and making higher contributions to local economies than other kinds of tourists. This resembles McKercher and du Cros' (2002) cultural tourism typology's 'highly motivated cultural tourists', to whom the destination culture is the main motivation to travel. According to Wu *et al.* (2014) serious indigenous tourists 'experience indigenous cultures in ways that better meet the principles of sustainable development', which is consistent with most of the definitions of indigenous tourism.

Representing Ethnicity in Tourism

In general, the issues of control and power are crucial in ethnic tourism and tourism–community relations. However, they are not only related to a matter of direct benefits from tourism, i.e. how the rules of 'exchange' and benefit sharing are defined and set in practice. In addition, Telfer and Sharpley (2007) have pointed out that ethnic tourism should respect local cultures, identities and traditions. Therefore, the ethnic people and communities that are used in and involved with tourism should be depicted in ways that are politically correct and socially acceptable (Saarinen, 1999, 2011). Therefore, the ways ethnic groups are illustrated and framed in place promotion, marketing and products are highly critical issues (Cohen, 1993; Hunter, 2011), especially in the context of sustainable tourism.

Depictions of destinations and related host cultures are crucial in tourism and how destinations aim to attract visitors (Echtner & Prasad, 2003). While there are many positive examples of socially sustainable ways of using and creating ethnic identities in tourism, a general tendency seems to be geared towards more problematic issues. Indeed, past research on how people are used in tourism and depicted as tourist attractions has been mainly critical (see Waitt, 1999). According to many authors, ethnic minorities and especially indigenous people are often represented in a primitive and exotic sense with reference to the colonial era, nostalgic, wild and heroic pasts, and early discoveries of Africa (see Bruner, 2002; van Eeden, 2006). This primitivisation has been a typical element in Western travel literature when representing indigenous cultures and peoples (see Wels, 2004). Edwards (1996) has further pointed out that the construction of the 'exotic other' makes indigenous people an integral part of the natural environment but does not necessarily suggest their power over it. Unfortunately, these kinds of visual depictions are somewhat typical for tourism promotion in southern Africa. For example, the Ovahimba (Himba) are a highly visible ethnic group in Namibian tourism marketing and are also used in a wider regional context (Saarinen & Niskala, 2009). The Ovahimba form an ethnic minority that lives in north-west Namibia and south-west Angola (Bollig & Heinemann, 2002). Traditionally they have been pastoralist, semi-nomadic people, with distinct traditions and cultural features. The Ovahimba female hairstyle, clothing and their tradition of using a mixture of ochre, butter fat and herbs to cover their skin are highly recognisable. The role of the Ovahimba in tourism promotion is often as a passive and posing object for tourists to gaze at (Saarinen, 2011, 2012; van Eeden, 2006). In addition, the visualisation of ethnicity often creates a highly gendered landscape; in the case of the Ovahimba, men or families are rarely depicted (Kanguma, 2000). This gendered imbalance in tourism marketing refers to the historically contingent modes of representing indigenous people as primitive, exotic and erotic objects (Edwards, 1996). In addition, these kinds of representations of local cultures create touristic expectations (see Rothfuss, 2000) that hosts often try to satisfy in order to stay attractive and receive benefits from visiting tourists. This 'governmentality', together with related tourism development practices, may also have an effect on how ethnic groups perceive their role and options in tourism and what possibilities they are given to participate in to benefit from tourism (Saarinen, 2011).

Cultural Villages: Time-Space Compressions in Ethnic Tourism

One interesting and evolving form of ethnic tourism in southern Africa is based on so-called cultural villages (van Veuren, 2001). In recent decades these villages have evolved in greater numbers in the region (see Saarinen, 2007; van Veuren, 2004). Together with urban-oriented township tours, cultural villages are currently perhaps the main form of cultural tourism in southern Africa. Van Veuren (2004: 140) has estimated that almost one-third of international tourists visit a cultural village during their stay in South Africa.

Cultural villages can be characterised as commercial and/or heritage sites where a particular time, place and culture are reconstructed in a contemporary context (Cameron, 1997: 136). They are specific attractions symbolising the way of living of local people (Timothy & Boyd, 2006). Typically, cultural villages aim to demonstrate traditional cultures through model homes, entertainment, stories, food, household activities, tools and clothing. In addition to displays and built structures, the activities and attractions related to cultural villages often include guided tours, cultural shows, exhibitions and craft workshops (Mbaiwa & Sakuze, 2009; Saarinen, 2007).

Motivations to establish cultural villages in southern Africa vary from highly touristic aims to cultural preservation and regional or local identity politics. In the South African context, van Veuren (2004: 141) has categorised cultural villages into three different types based on their ownership and funding structure:

(1) The first type of village – privately managed cultural villages – are mainly owned by non-local entrepreneurs representing other ethnic groups than the depicted one(s). These cultural villages are pure tourist products aiming to make revenue and profit for the business and owners.
(2) The second type of private sector-owned cultural village comprises 'native', i.e. local, entrepreneurs representing an ethnic group(s) depicted in villages. These villages are also tourist products but often involve a commitment to cultural conservation, education and local employment creation.
(3) The third type of cultural village is public sector driven. These are aiming to preserve elements of specific ethnic cultures and, thus, they are often part of wider national or regional identity politics. While tourism may play a role in their functions, these cultural villages rely heavily on government/public sector subsidies.

One of the most known and visited cultural villages in South Africa is the Lesedi (Plate 2.1). It is a tourist product, i.e. type one cultural village, located in Gauteng Province near the cities of Pretoria and Johannesburg. The village was established in 1995 and it is part of a multinational hotel chain: the Protea Hospitality Corporation (owned by the Marriott International), which is the largest hotel management group in Africa. In addition to accommodation and conference facilities, there are cultural programmes and tours offered through traditional homesteads, craft shops, storytelling and restaurant services. The different homesteads situated at the village include five different cultures (Basotho, Ndebele, Pedi, Xhosa and Zulu). The tour to these homesteads is followed by the performance that includes a set of

The Role of Local Cultures and Ethnicity in Tourism Development 25

Plate 2.1 The Lesedi Cultural Village, South Africa
Photo: Jarkko Saarinen.

traditional and modern dance shows, ranging from traditional 'ethnic' dances to the rhythms based on the black mining work communities during the apartheid era. As such, the Lesedi Cultural Village is a time-space compression which includes not only different cultures and regions in the same place ('A kaleidoscope of cultural colour') but also narratives and elements from the distant past and present (Saarinen, 2007).

In addition to Lesedi, there are over 30 other cultural villages in South Africa, such as Gaabo Motho Cultural Village (Gauteng), Sudwala and Shangana Cultural Villages (Mpumalanga) and Shakaland Zulu Cultural Village (KwaZulu-Natal). Other southern African countries also have cultural villages. In Namibia, for example, they depict Ovahimba (e.g. Purros Cultural Village in Kunene region) and Damara cultures (e.g. the Living Museum of the Damara in Kunene region). In Botswana, recent emphasis on the potential for cultural and heritage tourism development has resulted in the emergence of cultural villages (Saarinen *et al.*, 2014). The ethnic dimension is mainly based on a hegemonic Tswana culture but also the San (Basarwa) culture is utilised in cultural villages. In Botswana, cultural villages are industry-oriented, and the business development approach has been used by the government to encourage entrepreneurship. As a result, most villages, such as Bahurutse (Kweneng District) and Xaixai cultural villages (Ngamiland District), around the country are individually 'family' owned (Mbaiwa & Sakuze, 2009; Monare, 2013). In contrast to this, is the only cultural village in Swaziland – the Swazi Cultural Village (or Mantenga Cultural Village), which is public sector driven and owned by the state. The village's official name is 'Ligugu Lemaswati' – the pride of the Swazi People – Cultural Village, which demonstrates its role in identity politics and nation-making (Saarinen, 2007). Indeed, the founding objective of the village (in 1998) was to enable Swazi's from all corners of the country to reach out to it and maintain a positive interest in their cultural heritage as well as showing tourists their cultural achievements.

Concluding Remarks

Many southern African countries see the promotion of cultural tourism as a viable and economic strategy to develop tourism and integrate the industry into localities that tourists visit. As a result, many communities and ethnic groups are increasingly tied to tourism and its transnational socio-economic, cultural and political networks. This, together with a growing interest in cultural tourism among visiting tourists, has resulted in an increasing number of cultural tourism attractions, such as cultural villages, community-based tourism products and township tours in southern Africa. Many of these cultural sites are related to ethnicity and indigeneity, which have provided new employment and livelihood options for peripheral communities.

However, while there are many beneficial elements, tourism–community relations can also be challenging, especially when they involve ethnic minority groups and indigenous people. Therefore, many scholars have emphasised the need for sustainability in the context of ethnic and indigenous tourism development and governance (see Hinch & Butler, 2007). Indeed, for ethnic minorities the modern tourism industry can represent a major force for change, and while tourism can provide possibilities and new ways for local indigenous communities to live and maintain and value their culture, it can also create unwanted outcomes and images that are difficult or even impossible to reverse.

Therefore, in order to minimise the costs and maximise the benefits from tourism development, there should be a clear and firm understanding of the nature and limits of tourism in ethnic communities. Thus, there needs to be a recognition of what changes can be considered acceptable and how these changes occur and can be guided and controlled. This aim to think about the limits of growth in tourism and prioritise community views in tourism development refers to the principles of sustainable tourism and, especially, community-based approaches to sustainability (Saarinen, 2014). In community-based approaches, the host community and the benefits that it may gain from tourism are in a central position. This perspective underlines a need for research into the political economy and political ecology of tourism, which have been somewhat ignored issues in tourism studies (see Bianchi, 2004; Mosedale, 2011). These approaches could offer fruitful avenues for ethnic tourism studies and development as they can provide tools to analyse how potentially unequal power relations influence the level and nature of benefits from the utilisation of community resources (cultural and/or natural) in tourism (see Bryant & Bailey, 1997). Thus, together with an emphasis on sustainable tourism, the political economy and ecology could help us to understand and analyse what the locally desired conditions are for tourism development and its limits.

References

Ashley, C. and Roe, D. (2002) Making tourism work for the poor: Strategies and challenges in southern Africa. *Development Southern Africa* 19 (1), 61–82.
Barnard, A. (2006) Kalahari revisionism, Vienna and the 'indigenous peoples' debate. *Social Anthropology* 14 (1), 1–16.
Bianchi, R. (2004) Tourism restructuring and the politics of sustainability: A critical view from the European periphery (The Canary Islands). *Journal of Sustainable Tourism* 12 (6), 495–529.
Bollig, M. and Heinemann, H. (2002) Nomadic savages, ochre people and heroic herders: Visual presentation of the Himba of Namibia's Kaokoland. *Visual Anthropology* 15, 267–312.
Bruner, E.M. (2002) The representation of African pastoralists: A commentary. *Visual Anthropology* 15 (3), 387–392.
Bryant, R. and Bailey, S. (1997) *Third World Political Ecology*. New York: Routledge.
Cameron, G. (1997) Curricular discourses at a southeast Asia cultural village. In V. Smith (ed.) *Hosts and Guests: The Anthropology of Tourism* (pp. 136–149). Philadelphia: University of Pennsylvania Press.
Cohen, E. (1993) The study of touristic images of native people: Mitigating the stereotype for a stereotype. In D.G. Pearce and R.W. Butler (eds) *Tourism Research: Critiques and Challenges* (pp. 36–69). London: Routledge.
Echtner, C.M. and Prasad, P. (2003) The context of Third World tourism marketing. *Annals of Tourism Research* 30, 660–682.
Edwards, E. (1996) Postcards: Greetings from another world. In T. Selwyn (ed.) *The Tourist Image. Myths and Myth Making in Tourism* (pp. 61–81). Chichester: Wiley & Sons.
Greene, S. (2004) Indigenous people incorporated? *Current Anthropology* 45, 211–237.
Grunewald, R. (2002) Tourism and cultural revival. *Annals of Tourism Research* 29, 1004–1021.

Hinch, T. and Butler, R. (1996) Indigenous tourism: A common ground for discussion. In R. Butler and T. Hinch (eds) *Tourism and Indigenous Peoples* (pp. 1–12). London: International Thomson Business Press.

Hinch, T. and Butler, R. (2007) Indigenous tourism: Revisiting common ground. In R. Butler and T. Hinch (eds) *Tourism and Indigenous Peoples* (pp. 3–19). Oxford: Butterworth-Heinemann.

Hunter, W.C. (2011) Rukai indigenous tourism: Representations, cultural identity and Q method. *Tourism Management* 32, 335–348.

ILO (International Labour Organization) (1989) *Indigenous and Tribal Peoples Convention No. 169*. Geneva: General Conference of the International Labour Office.

Ivanovic, M. (2008) *Cultural Tourism*. Cape Town: Juta.

Ivanovic, M. and Saayman, A. (2013) South Africa calling cultural tourists. *African Journal for Physical, Health Education, Recreation and Dance* 19 (2), 138–154.

Kanguma, B. (2000) Constructing Himba: The tourist gaze. In G. Miescher and D. Henrichsen (eds) *New Notes on Kaoko* (pp. 129–132). Basel: Basler.

Kuper, A. (2003) The return of the native. *Current Anthropology* 44, 389–402.

Lapeyre, R. (2010) Community-based tourism as a sustainable solution to maximise impacts locally? The Tsiseb Conservancy case, Namibia. *Development Southern Africa* 27 (5), 757–772.

Li, W.J. (2006) Community decision making: Participation in development. *Annals of Tourism Research* 33 (1), 132–143.

Manwa, H. (2003) Wildlife-based tourism, ecology and sustainability: A tug-of-war among competing interests in Zimbabwe. *The Journal of Tourism Studies* 14 (2), 45–54.

Mbaiwa, J.E. and Sakuze, L.K. (2009) Cultural tourism and livelihood diversification: The case of Gcwihaba Caves and XaiXai village in the Okavango Delta, Botswana. *Journal of Tourism and Cultural Change* 7 (1), 61–75.

McKercher, B. and du Cros, H. (2002) *Cultural Tourism: The Partnership between Tourism and Cultural Heritage Management*. New York: The Haworth Hospitality Press.

Monare, M. (2013) Local people's attitudes and perceptions towards benefits and costs of cultural tourism: Case studies of Bahurutshe and Motse cultural villages, Botswana. Master's Thesis, University of Botswana.

Mosedale, J. (ed.) (2011) *Political Economy of Tourism*. Oxon: Routledge.

Nijman, J. (1999) Cultural globalization and the identity of place: The reconstruction of Amsterdam. *Cultural Geographies* 6 (2), 146–164.

Poon, A. (1993) *Tourism, Technology and Competitive Strategies*. Wallingford: CAB International.

Richards, G. (ed.) (1997) *Cultural Tourism in Europe*. Wallingford: CAB International.

Rogerson, C. (2015) Tourism and regional development: The case of South Africa's distressed areas. *Development Southern Africa*. DOI: 10.1080/0376835X.2015.1010713.

Rogerson, C. and Visser, G. (2004) Tourism and development in post-apartheid South Africa: A ten-year review. In C.M. Rogerson and G. Visser (eds) *Tourism and Development Issues in Contemporary South Africa*. Pretoria: Africa Institute of South Africa.

Rothfuss, E. (2000) Ethnic tourism in Kaoko: Expectations, frustrations and trend in a post-colonial business. In G. Miescher and D. Henrichsen (eds) *New Notes on Kaoko* (pp. 133–159). Basel: Basler.

Ryan, C. (2005) Introduction: Tourist-host nexus – Research considerations. In C. Ryan and M. Aicken (eds) *Indigenous Tourism: The Commodification and Management of Culture* (pp. 1–11). Oxford: Elsevier.

Ryan, C., Hughes, K. and Chirgwin, S. (2000) The gaze, spectacle and ecotourism. *Annals of Tourism Research* 27 (1), 148–163.
Saarinen, J. (1999) Representations of indigeneity: Sami culture in the discourses of tourism. In P.M. Sant and J.N. Brown (eds) *Indigeneity: Constructions and Re/presentations* (pp. 231–249). New York: Nova Science Publishers.
Saarinen, J. (2007) Cultural tourism, local communities and representations of authenticity: The case of Lesedi and Swazi cultural villages in Southern Africa. In B. Wishitemi, A. Spenceley and H. Wels (eds) *Culture and Community: Tourism Studies in Eastern and Southern Africa* (pp. 140–154). Amsterdam: Rozenberg.
Saarinen, J. (2011) Tourism, indigenous people and the challenge of development: The representations of Ovahimbas in tourism promotion and community perceptions towards tourism. *Tourism Analysis* 16 (1), 31–42.
Saarinen, J. (2012) Tourism development and local communities: The direct benefits of tourism to OvaHimba communities in the Kaokoland, North-West Namibia. *Tourism Review International* 15, 149–157.
Saarinen, J. (2013) Indigenous tourism and the challenge of sustainability. In M. Smith and G. Richards (eds) *Routledge Handbook of Cultural Tourism* (pp. 220–226). London: Routledge.
Saarinen, J. (2014) Critical sustainability: Setting the limits to growth and responsibility in tourism. *Sustainability* 6 (11), 1–17.
Saarinen, J., Moswete, N. and Monare, M.J. (2014) Cultural tourism: New opportunities for diversifying the tourism industry in Botswana. *Bulletin of Geography. Socio-Economic Series* 26, 7–18.
Saarinen, J. and Niskala, M. (2009) Local culture and regional development: The role of OvaHimba in Namibian tourism. In P. Hottola (ed.) *Tourism Strategies and Local Responses in Southern Africa* (pp. 61–72). Wallingford: CABI Publishing.
Santos, C. and Yan, G. (2008) Representational politics in Chinatown: The ethnic other. *Annals of Tourism Research* 35 (4), 879–899.
Saugestad, S. (2001) *The Inconvenient Indigenous*. Uppsala: The Nordic Africa Institute.
Sinclair, D. (2003) Developing indigenous tourism: Challenges for the Guianas. *International Journal of Contemporary Hospitality Management* 15, 140–146.
Smith, M. (2003) *Issues in Cultural Tourism Studies*. London: Routledge.
Smith, M. and Richards, G. (eds) (2013) *The Routledge Handbook of Cultural Tourism*. New York; London: Routledge.
Smith, V.L. (ed.) (1989) *The Host and Guests: The Anthropology of Tourism*. Philadelphia: University of Pennsylvania Press.
Smith, V.L. (1996) Indigenous tourism: The four Hs. In R. Butler and T. Hinch (eds) *Tourism and Indigenous Peoples* (pp. 283–307). London: International Thomson Business Press.
Stronza, A. (2007) The economic promise of ecotourism for conservation. *Journal of Ecotourism* 6, 210–230.
Telfer, D. and Sharpley, R. (2007) *Tourism and Development in the Developing World*. London: Routledge.
Timothy, D.J. and Boyd, S.W. (2006) Heritage tourism in the 21st century: Valued traditions and new perspectives. *Journal of Heritage Tourism* 1 (1), 1–16.
United Nations (2004) The concept of indigenous indigenous peoples. New York: Department of Economic and Social Affairs, United Nations. See www.un.org/esa/socdev/unpfii/documents/workshop_data_background.doc (accessed 12 December 2011).

United Nations (2007) *The United Nations Declaration on the Rights of Indigenous Peoples*. New York: The United Nations. See www.un.org/esa/socdev/unpfii/documents/DRIPS_en.pdf (accessed 10 November 2011).

Urry, J. (1990) *The Tourist Gaze*. London: Sage Publications.

van Eeden, J. (2006) Land Rover and colonial-style adventure. *International Feminist Journal of Politics* 8, 343–369.

van Veuren, E.J. (2001) Transforming cultural villages in the spatial development initiatives of South Africa. *South African Geographical Journal* 83 (2), 137–148.

van Veuren, E.J. (2004) Cultural village tourism in South Africa: Capitalizing on indigenous culture. In C.M. Rogerson and G. Visser (eds) *Tourism and Development Issues in Contemporary South Africa* (pp. 139–160). Pretoria: Africa Institute of South Africa.

Waitt, G. (1999) Naturalising the 'primitive': A critique of marketing Australia's indigenous peoples as 'hunter-gatherers'. *Tourism Geographies* 1 (1), 142–163.

Wels, H. (2004) About romance and reality: Popular European imagery in postcolonial tourism in Southern Africa. In C.M. Hall and H. Tucker (eds) *Tourism and Postcolonialism: Contested Discourses, Identities and Representation* (pp. 76–94). London: Routledge.

Wu, T.-C., Wall, G. and Tsou, L.-Y. (2014) Serious tourists: A proposition for sustainable indigenous tourism. *Current Issues in Tourism*. DOI: 10.1080/13683500.2014.970143.

Yang, L. (2011) Ethnic tourism and cultural representation. *Annals of Tourism Research* 38 (2), 561–585.

Yang, L. and G. Wall (2009) Ethnic tourism: A framework and an application. *Tourism Management* 30, 559–570.

Yang, L., Wall, G. and Smith, S. (2008) Ethnic tourism development: Chinese government perspectives. *Annals of Tourism Research* 35 (3), 751–771.

3 Integrating Indigenous Knowledge in the Development of Cultural Tourism in Lesotho

Tsitso Monaheng

Introduction

There are indications that cultural tourism is growing faster than other forms of tourism, particularly in developing countries, and that it has the potential to contribute towards alleviating poverty and promoting community economic development (Timothy & Nyaupane, 2009). Earlier, Viljoen and Tlabela (2007) voiced a similar view by suggesting that cultural tourism is considered to be a way in which disadvantaged communities and persons are able to benefit from tourism by using their culture as an attraction. While recognising the growth in cultural tourism, Schouten (2007: 26) expresses scepticism about the impact of this growth, compared to what he refers to as 'traditional forms of tourism', given that it was starting from a very low base.

The foregoing highlights the fact that the importance of cultural tourism should, at least in part, be linked to the contribution that it can make towards enhancing the economic well-being of people in the Third World. However, Richards (2007: 1) provides a broader perspective on the significance of cultural tourism by observing that: 'UNESCO promotes cultural tourism as a means of preserving world heritage, the European Commission supports cultural tourism as a major industry, and the newly emerging nation-states of Africa and Central Europe see it as a support for national identity. In many parts of the world it has become a vital means of economic support for traditional activities and local creativity'.

It is therefore essential to bear in mind that in looking at cultural tourism within the context of development, cognisance should be taken of the multiplicity of objectives that it can be used to achieve, given that development itself is a multifaceted phenomenon.

Indigenous Knowledge Systems (IKS)

Indigenous knowledge (IK) is seen as knowledge found in local communities. It is knowledge which the people have used for many years to organise their lives and to survive; it is passed down from generation to generation mainly by word of mouth (Kaya & Masoga, in Dweba & Mearns, 2011: 264). IK reflects the local people's approach to, and understanding of, issues relating to environmental protection, food security and technology, agriculture, social welfare, conflict resolution and medicine (Odora Hoppers, in Dweba & Mearns, 2011). From Nel's (2006: 99) perspective, indigenous knowledge encompasses the belief systems, art work, customs, cultural practices and values, production technologies as well as 'ways of knowing and sharing in terms of which communities have survived for centuries'. Dei et al. (2000) suggest that indigenous knowledge embodies the totality of the people's world view. It helps them to organise and regulate not only their ways of making a living but also their philosophies of making sense of the world around them, and forms the basis of the decisions they make in their daily lives.

IKS and Tourism

There are close interconnections between cultural and heritage tourism, sustainable tourism and community-based tourism. Within the context of what they define as community-based 'eco-cultural tourism', Turner et al. (2012) have identified the main objectives to be environmental protection, socio-economic well-being and the preservation of cultural values. These goals have also been acknowledged by Walter (2009) and Lukhele and Mearns (2013), in the context of 'community-based ecotourism' and 'community-based tourism', respectively. Walter (2009) as well as Walter and Reimer (2012) have also demonstrated that there is a close relationship between indigenous knowledge and the attainment of the objectives of cultural tourism.

Environmental protection

The significance of the role of indigenous knowledge in nature conservation has been noted by different writers, and numerous calls have been made for IK to be integrated in efforts directed at environmental protection. Marie et al. (2009), for example, argue that the success of conservation projects depends on three factors, namely: (a) the participation of the local people in determining the objectives of conservation; (b) recognition of the value of IK and (c) creating potential sources of income through tourism. They maintain that local populations have, over many generations, developed knowledge that has enabled them 'to maintain the sustainability of their farming systems and handle the issue of resource renewal'. Popova (2014), Raymond et al. (2010), as well as Taylor and de Loë (2012), have, similarly, demonstrated that local communities have cultural rules, norms and knowledge which they have used, and continue to use, in their interaction with the natural environment.

There are, nevertheless, problems militating against the integration of indigenous knowledge in conservation and natural resource management projects. The problems have also been identified in other areas of IK. One is that there is reluctance on the part of some role players to accept the validity of IK. Unequal relations of power between this group, on the one hand, and supporters of indigenous knowledge systems, on the other, often work against the acceptance of this from of knowledge (Gaillard & Mercer, 2013; Taylor & de Loë, 2012). Another problem, which is also related to disproportionate power relations, relates to loss of intellectual property rights by the local people over their indigenous knowledge products. These rights are often appropriated by scientists and researchers or the institutions to which they are attached; this limits the extent to which the locals are able to benefit from their IK (Popova, 2014; Prasad, 1998).

Economic and livelihood activities

Because of greater dependence on the natural environment for their livelihoods, rural communities exhibit greater reliance on IK, to deal with the challenges they face in earning a living (Teffo, 2013). In their study of communities living in the Himalaya Mountains of India, Samal et al. (2004) conclude that the people have developed agricultural practices that enable them to cope with their environment. They also have knowledge of wild plants – those that are used for food and those used for medicinal purposes. In a study undertaken in a village in the Eastern Cape Province of South Africa, Dweba and Mearns (2011) similarly report on the knowledge of the people about traditional vegetables and medicinal herbs. Grice et al. (2012)

also allude to the role of IK in promoting sustainable economic well-being of local communities.

Some tourism products enable local communities to earn a livelihood by sharing their IK with other people. Walter (2009) together with Walter and Reimer (2012) report on cases in Thailand and Cambodia where tourists participate in local livelihood activities as a way of learning about the cultures of the people and experiencing the sustainable use of flora and fauna in the area. They partake in hunting and fishing expeditions, gather wild fruits and vegetables, engage in agricultural activities and in food preparation. Homestays serve as a form of cultural exchange in which tourists learn the local culture while also imparting some aspects of their culture to their hosts.

Box 3.1 Cultural products: A driver for informal sector business tourism in southern Africa

Christian M. Rogerson

In a seminal study, Rob Davidson (1994: 1) defines business tourism as concerned 'with people travelling for purposes which are related to their work' and considers that 'it represents one of the oldest forms of tourism'. Northern scholarship on business tourism centres on travel for work purposes which is viewed as including the categories of general business travel, meetings, exhibitions and incentive travel (MICE tourism). These practices of business tourism are considered a vital and expanding element of tourism in Western economies with many countries and cities competing aggressively to attract MICE tourism through providing convention and exhibition facilities as well encouraging a network of modern accommodation facilities geared to the needs of the business travellers.

In the global south business tourism is equally a phenomenon of considerable importance, albeit largely one that has been overlooked by African tourism scholars (Rogerson & Rogerson, 2011). One dimension of business tourism in sub-Saharan Africa which parallels trends in the global north is the expansion of various kinds of formal sector business travel that relate to international or local business meetings or conferences (Rogerson, 2005). But, alongside this form of business tourism there is another much less well understood economy of *informal sector* business tourism which has both international and domestic manifestations. Across sub-Saharan Africa, informal sector business tourism is a substantial element, if not the largest numerical constituent of business travel and tourism. Mitchell and Ashley (2010) maintain that the scale of domestic business tourism in sub-Saharan Africa is highly significant in terms of total expenditure as well as its pro-poor impacts.

The aim here is to draw attention to the role of cultural products as drivers of informal sector business tourism in southern Africa. Cultural products cover a broad spectrum of goods and would include, for example, traditional medicines, herbs, crafts, ethnic products and handcrafted goods which historically have been produced to serve as functional items within local communities (Kumphai, 2006). In many parts of the global south certain traditional cultural products have been commercialised and transformed into cottage industries with the making of artisanal products a vital income source for rural communities. In a small number of instances the commercialisation of traditional cultural products (such as certain handcrafts) has reached a level where cultural products are exported. In southern Africa the economic role of cultural products remains of vital importance in both urban and rural environments. For example, in Johannesburg or Durban, two of Africa's most modernised and economically vibrant cities, the role of certain traditional cultural products remains undiminished. Notwithstanding the existence of modern health care facilities, both cities contain, for example, an economy of traditional herbalism which is associated with the continued role of traditional medicine in urban communities. In smaller cities of southern Africa the persistent use of cultural products is widely recognised in, for example, Blantyre, Mbabane or Maseru. Other cultural products are traded as well. For example, a range of handcrafted goods produced by artisanal craftsmen (and women) in rural areas are widely sold in urban markets both to international and local tourists as well as to domestic consumers.

What is important to recognise is the geographical connection between the production of such cultural products, which primarily occurs in (often remote) rural areas, and the selling or distribution of these goods which takes place in urban centres. The trade in traditional cultural products has been an historical driver for the establishment and growth of forms of informal sector business tourism. Indeed, for South Africa, Cunningham (1991) and Dauskardt (1991) provide evidence of the complex web of rural–urban connections which have resulted in informal sector business mobilities as traders bring goods into cities for sale either directly to consumers at markets or to urban wholesalers or

(continued)

Box 3.1 Cultural products: A driver for informal sector business tourism in southern Africa (*continued*)

retailers. Among others, the work of Street and Prinsloo (2013) attests to the continued relevance of this trade. Although anecdotal evidence exists of the phenomenon of informal sector domestic business tourism occurring in several different African countries (Rogerson, 2012), the only documented case is of Maseru, Lesotho's capital city. The nature of informal sector business travellers in Maseru includes rural producers of craft goods (such as traditional woven hats), the makers of traditional weapons and the circulation of herbal traders (Rogerson & Letsie, 2013).

Overall, in southern Africa cultural products are drivers of informal sector forms of business tourism that connect rural and urban communities. The character of these domestic business tourists trading cultural products departs radically from conventional 'northern' definitions of business tourists. This demonstrates that the nature of the African tourist departs from westernised notions of the 'business traveller' and of what constitutes business tourism in northern scholarship.

References

Cunningham, A. (1991) The herbal medicine trade: Resource depletion and environmental management for a hidden economy. In E. Preston-Whyte and C. Rogerson (eds) *South Africa's Informal Economy* (pp. 196–206). Cape Town: Oxford University Press.

Dauskardt, R. (1991) 'Urban herbalism': The restructuring of informal survival in Johannesburg. In E. Preston-Whyte and C. Rogerson (eds) *South Africa's Informal Economy* (pp. 87–100). Cape Town: Oxford University Press.

Davidson, R. (1994) *Business Travel*. London: Pitman.

Kumphai, P. (2006) Cultural products: Definition and website evaluation. MSc dissertation, Oklahoma State University.

Mitchell, J. and Ashley, C. (2010) *Tourism and Poverty Reduction: Pathways to Prosperity*. London: Earthscan.

Rogerson, C.M. (2005) Conference and exhibition tourism in the developing world: The South African experience. *Urban Forum* 16, 176–195.

Rogerson, C.M. (2012) The tourism-development nexus in sub-Saharan Africa: Progress and prospects. *Africa Insight* 42 (2), 28–45.

Rogerson, C.M. and Letsie, T. (2013) Informal sector business tourism in the global South: Evidence from Maseru, Lesotho. *Urban Forum* 24 (4), 485–502.

Rogerson, C.M. and Rogerson, J.M. (2011) Tourism research within the Southern African Development Community: Production and consumption in academic journals, 2000–2010. *Tourism Review International* 15 (1–2), 213–222.

Street, R.A. and Prinsloo, G. (2013) Commercially important medicinal plants in South Africa: A review. *Journal of Chemistry* 2013, 1–16. DOI/10.1155/2013/205048.

Preservation of culture

The interrelationships among the objectives of nature conservation, socio-economic development and maintenance of cultural integrity have been highlighted by writers such as Marie *et al.* (2009), Ndlovu (2004) and Turner *et al.* (2012). Especially in rural and traditional communities, cultural norms and practices form the basis of economic activities and influence the way in which people relate to their environment. In turn, experiences gained from earning a livelihood and interacting with the natural environment shape the nature of their cultural values.

Various writers examine the relationship between IK and different elements of culture. Harrison and Papa (2005: 57) highlight this link by reporting on how an indigenous community in New Zealand decided that primary and secondary school children should be taught in their language not only to improve 'student achievement' but also for the purpose of 'maintaining the Maori language, [and] providing children with knowledge and confidence in their heritage'. Willox *et al.* (2013) talk about oral storytelling as a medium used to remember, connect with and share a people's heritage. They also hail the use of digital technology (e.g. photographs, video, text, websites, social networking sites, email) as a way of preserving and sharing IK.

Seema (2012: 128) alludes to the importance of the spoken word by arguing that IK among the Basotho (citizens of Lesotho, also found in South Africa – mainly in the Free State Province) is expressed through 'stories, legends, folklore, rituals, songs and proverbs'. About proverbs, Seema (2012: 129) notes that they 'embody the collective wisdom and resourcefulness for the development of the community'. With regard to African music, Joseph (2005) contends that not only does it serve an aesthetic purpose but that it also promotes African cultural values. Drawing from Westerlund and Miller, Joseph (2005) suggests that African music reinforces the African way of life of sharing as well as community participation and cooperation, as it is intended for the audience to join in and become part of the action.

Questions have been raised about whether or not cultural tourism serves to promote or to compromise the authenticity of the culture of the host community. However, it is also important to realise that the authenticity of different aspects of culture is not always defined in terms of originality or the length of time they have existed. Due to internal social dynamics and external influences, including the need to cater for the preferences of tourists, local cultures are continuously undergoing change (Schouten, 2007; Sigala & Leslie, 2005; Timothy & Boyd, 2003). Even where changes to culture are influenced by tourists, Smith (2009) maintains that it is important that local people accept them. Sofield (in Timothy & Nyaupane, 2009: 84–85) expresses a similar view by recording that 'each generation redefines its heritage in response to new understandings, new experiences and new inputs from an ever-increasing range of contacts from outside'. Schouten (2007: 34) also says, 'Heritage is re-evaluated by each generation anew; some aspects are added to it and others may be fading away.' The dynamism of IK has equally been acknowledged, as it evolves by interacting with other forms of knowledge (Briggs, 2005; Nel, 2006; Teffo, 2013).

Cultural Tourism in Lesotho

Tourism has been identified as a key element of Lesotho's development strategy, directed specifically at poverty alleviation (UNDP/UNWTO 2006). The government considers it to be an important sector 'to drive growth and employment because of its labour-intensive nature and its potential to raise income in rural areas' (Lesotho Government, 2012: 102). Tourism development has also been viewed as a basis for community development (Lesotho Review, 2004). Similarly, Mashinini (2003) has determined that the following are among the main objectives of tourism policy in Lesotho: the creation of employment, the promotion of rural tourism to reduce rural–urban inequality, encouraging community participation in tourism and the development of sustainable tourism.

The strategic plan for tourism development in Lesotho (Lesotho Government, 2007: 6) also reveals that while the country has numerous tourism resources their potential has not been adequately tapped. In part, these resources have been described in the following manner:

- The unique natural environment including the mountain scenery and scenic routes, topography, flora and fauna.
- The rich Basotho culture and lifestyle that is intertwined with the physical environment and altitude – ponies as transport, stone architecture, blankets as protective wear, migratory grazing and herdboy culture, Basotho music and rich cultural traditions.
- A rich heritage including the Mountain Kingdom 'story', Thaba-Bosiu (meaning, the mountain which grows at night), Liphofung caves, San rock art, and dinosaur footprints.

Consequently, the forms of tourism to which Lesotho is paying particular attention are variously described as ecotourism, heritage and cultural tourism, sustainable tourism and community-based tourism (Lesotho Government, 2007; Lesotho Review, 2004, 2010, 2014). It has already been demonstrated that there are close interrelationships among these types of tourism.

Conservation and Ecotourism

One of the most important tourism development initiatives in Lesotho is the Maluti-Drankensberg Transfrontier Programme (MDTP), which is a joint venture between Lesotho and South Africa. The MDTP is intended to promote cooperation in conservation and sustainable development initiatives in the project area and to encourage community involvement in nature-based tourism (Büscher, 2010; Cain, 2009). As part of this programme, between 2005 and 2006 the Government of Lesotho commissioned a cultural heritage survey; it identified numerous previously unrecorded heritage sites – including, hunter-gatherer sites, rock art sites, historical sites and living heritage sites (Cain, 2009).

Lesotho has three national parks, namely, the Tsehlanyane National Park, the Sehlabathebe National Park and the Bokong Nature Reserve. The Sehlabathebe National Park was declared a World Heritage Site in 2013, as an extension of South Africa's Ukhahlamba–Drakensberg Park (a World Heritage Site in its own right). Together the two parks are now known as the Maloti-Drakensberg Park, Lesotho/South Africa (Lesotho Review, 2014). This area forms part of the Maluti-Drankensberg transfrontier conservation initiative between the two countries, referred to above.

While accepting that national parks 'are the best way to conserve biodiversity as they protect species in their natural habitats', Mofokeng and Ge (2009) maintain that in Lesotho national parks alone cannot save endangered species. Partly, they attribute this to lack of information about the need for conservation among adjacent communities, but mainly to poor enforcement of laws. Citing the example of the spiral aloe (*Aloe polyphylla*), the national flower of Lesotho as well as a medicinal plant, they note that people tend to harvest plant species at will in the wild. They, therefore, suggest that the establishment of botanical gardens, which has only recently begun to take root in Lesotho, is the right way to go. The Katse Botanical Gardens, for example, were established to preserve rare and endangered plant species, including the spiral aloe. The Katse Dam was built on land where those plants thrived.

Another measure used to ensure the success of conservation efforts has been the promotion of community cooperation through the establishment of Community Conservation Forums (Lesotho Review, 2002). It is important also to realise that community participation has not only been encouraged to achieve sustainable management and use of resources, it has also been seen as a vehicle enabling communities to benefit from protected areas through access to employment opportunities, entrepreneurial and business opportunities as well as having a share of the profits from the ventures through community levies (Lesotho Review, 2002; UNDP/UNWTO, 2006).

Cultural Preservation and Display

Notable examples of efforts made to display and preserve cultural heritage in Lesotho are the Thaba-Bosiu Cultural Village, the Morija Arts and Cultural Festival, the Morija Museum & Archives and Moshoeshoe Day. The Thaba-Bosiu Cultural Village is intended to showcase different aspects of Basotho culture. It was constructed using original materials and methods. It is composed of 41 huts, a museum/interpretation centre, ceremonial spaces, a live-performance amphitheater, handicraft outlets and botanical gardens. In addition, it has 41 self-catering chalets and a conference centre (Lesotho Review, 2014). Cultural villages are also used in South Africa as repositories of culture and IK (Mearns, 2007; Viljoen & Tlabela, 2007).

The Morija Arts and Cultural Festival is held annually at the end of September or beginning of October, over a period of about four days. As Lesotho's premier cultural event, it focuses on keeping various aspects of Basotho culture alive. 'Attractions include arts and crafts, as well as cultural groups and performers – both traditional and modern – of music, dance, comedy and poetry' (Lesotho Review, 2014: 46). In Morija there is also the Morija Museum & Archives, which houses archival material and museum collections of national significance. Moshoeshoe Day is held annually in March. It commemorates the life of the founder of the Basotho Nation, King Moshoeshoe I, and it is a public holiday. Cultural performances also take place on this day.

The preservation of cultural heritage resources is a strategic objective of the Lesotho government. Measures to achieve this goal include the protection of heritage sites; proper management of museums and archives; encouraging documentation through art, writing, film and other means; and supporting the development of arts and crafts (Lesotho Government, 2012).

Community-based Tourism and IKS

Many examples of community-based tourism ventures have been initiated in different parts of Lesotho. They are based on various combinations of activities, but generally they cover the following areas: pony trekking, guided tours, homestays, handicraft centres, children's choirs and bands playing homemade instruments, local museums, visits to traditional healers, bird watching, trout fishing, visits to rock painting sites, waterfalls, dinosaur footprints and historic sites, traditional music and dances.

In a study based on the Malealea Lodge and Pony Trek Centre (a community-based ecotourism venture) in Lesotho, Mearns (2011) addresses issues of sustainability. He uses the framework that links (a) community issues and interests to social sustainability, (b) needs of the tourism industry to economic sustainability and (c) concerns about nature conservation to environmental sustainability. The three-fold classification of elements of sustainability in tourism is also used by writers such as Choi and Sirakaya (2006), Saarinen (2006, 2009) and Telfer and Sharpley (2008). Although Mearns (2011: 141) presents one of the indicators of social sustainability as 'cultural appreciation and conservation', the study found that 'there was a lower level of agreement that good souvenirs and crafts were available', which prompted the recommendation that 'further investigation into the types of souvenirs tourists actually want needs to be undertaken' (Mearns, 2011: 144). Clearly, this has implications for cultural authenticity and shows the potential impact of tourism on the culture of the Basotho.

Pheto-Moeti's (2005) study of the seshoeshoe dress as a source of cultural identity among Basotho women demonstrates the interplay between sustainability and authenticity of culture as well as IK (roughly translated the word seshoeshoe means 'of the Bashoeshoe'. Basotho are sometimes called Bashoeshoe, after the founder of the nation – King Moshoeshoe I. Pheto-Moeti reports that the original seshoeshoe has undergone changes due to influences from Western and other African cultures as well as technology (sewing machines able to make decorative stiches and embroidery). Her conclusions are that, although new ideas about the seshoeshoe are welcome, dressmakers should not depart too much from the original pattern as variations seem only to be acceptable to a certain extent. Also that, for cultural festivities, only the traditional design should be used. In a study of the Kome Cave Village, a heritage site and an important tourist attraction in Lesotho, Maanela (2008) highlights the need to balance the different aspects of sustainability, by emphasising the need to consider the interests of the community.

It is not always the case, however, that communities are aware of the possible negative impact of tourism on their culture. This is illustrated by Manwa's (2012) study of how some communities in Lesotho perceive the impact of tourism. In one of the three villages affected, the study found that tourism was viewed positively. This was partly because residents felt that it would revitalise traditional knowledge in the making of handicrafts as older people would see the benefit of transferring their skills to younger generations. Another point worthy of noting is made by Tanga and Maliehe (2011) in relation to the Malealea project already referred to. They report that women play a significant role in handicraft production, which they recognise as being based on IK. This underscores the role played by women in preserving culture and IK in Lesotho. Likewise, Nedelea and Okech (2008) have acknowledged the place of rural women as repositories of IK relating to handicrafts in South Africa.

Women's IK in Lesotho, especially rural women, extends to other areas as well. Prasad (1998) has noted that they have knowledge of plants that are used for food and as home remedies for some ailments. Prasad, nevertheless, also observed that this knowledge was disappearing partly due to competition from imported vegetables and Western medicine, the decline of the oral tradition and because it 'is associated with poverty and low status instead of being celebrated as a triumph of scientific excellence and mastery' (Prasad, 1998: 84). In Zimbabwe, Shava (2005) likewise witnessed the role of women as holders of IK about food plants, although it is also disappearing. Like Prasad in the case of Lesotho, he warned against the danger of intellectual rights to this knowledge being appropriated by others.

Mokuku (2012) explains an approach through which IK can be used to teach environmental education in schools through Lesotho's Education for Sustainable Development (ESD) programme. He uses the concept Lehae-La-Rona (Our Home) which he derives from Basotho's understanding of the interdependence between people, other animal species and plants. Mokuku illustrates this interrelationship by reporting that in an earlier research in the highlands of Lesotho they found that 'certain animals communicate essential messages or convey blessings to humans by performing certain movements or through encounter, and some plants and animals are accorded respect and reverence due to the powers and awe they embody' (Mokuku, 2012: 3). The concept of Lehae-La-Rona shows how cultural values and IK can be used to promote environmental sustainability in Lesotho. Indeed, it can serve not only to teach environmental education in schools but also to sensitise communities about preserving their heritage within the parameters of community-based eco-cultural tourism.

Conclusion

There exists a symbiotic relationship between culture and IK. Through IK people maintain and reproduce their culture. At the same time culture is an expression of a people's IK. Therefore, the two should be seen as opposite sides of the same coin. The concept of sustainability is a useful analytical tool for understanding the role of IK in cultural tourism. An appreciation of the social elements of sustainability helps to draw attention to the fact that the interests of the communities affected by tourism are not necessarily served by a static culture. Both IK and culture are constantly evolving, in response to internal social and economic dynamics as well as external influences.

It is, nevertheless, important that changes to culture are acceptable to the owners of the culture. This understanding leads to the realisation that the authenticity of culture is socially determined, as people interact with one another and accommodate different views and interests. Furthermore, it highlights the need to look at cultural tourism and IK within the context of a people-centred form of development, in which people's participation and empowerment play a central role. Genuine empowerment will ensure that the interests of the community affected by tourism are not overshadowed by the demands of the industry or concerns for environmental conservation. Also, more should be done to acknowledge the role of women in preserving cultural and IK resources by removing the impediments which inhibit their full participation in development.

References

Briggs, J. (2005) The use of indigenous knowledge in development: Problems and challenges. *Progress in Development Studies* 5 (2), 99–114.

Büscher, B. (2010) Derivative nature: Interrogating the value of conservation in 'Boundless Southern Africa'. *Third World Quarterly* 31 (2), 259–276.

Cain, C.R. (2009) Cultural heritage survey of Lesotho for the Maloti-Drakensberg Transfrontier Project: Palaeontology, archaeology, history and heritage management. *The South African Archaeological Bulletin* 64 (189), 33–44.

Choi, H.C. and Sirakaya, E. (2006) Sustainability indicators for managing community tourism. *Tourism Management* 27 (6), 1274–1289.

Dei, G.J., Hall, B.L. and Rosenberg, D.G. (2000) Introduction. In G.J. Dei, B.L. Hall and D.G. Rosenberg (eds) *Indigenous Knowledges in Global Contexts: Multiple Readings of Our World* (pp. 3–20). Toronto: University of Toronto Press.

Dweba, T.P. and Mearns, M.A. (2011) Conserving indigenous knowledge as the key to the current and future use of traditional vegetables. *International Journal of Information Management* 31 (6), 564–571.

Gaillard, J.C. and Mercer, J. (2013) From knowledge to action: Bridging gaps in disaster risk reduction. *Progress in Human Geography* 37 (1), 93–114.

Grice, A.C., Cassady, J. and Nicolas, D. (2012) Indigenous and non-indigenous knowledge and values combine to support management of Nywaigi lands in the Queensland coastal tropics. *Ecological Management & Restoration* 13 (1), 93–97.

Harrison, B. and Papa, R. (2005) The development of an indigenous knowledge programme in a New Zealand Maori-language immersion school. *Anthropology and Education Quarterly* 36 (1), 57–72.

Joseph, D. (2005) Localising indigenous knowledge systems down under: Sharing different worlds with one voice. *Indilinga – African Journal of Indigenous Knowledge Systems* 4 (1), 295–305.
Lesotho Government (2007) *A Strategic Plan for Tourism Development in Lesotho.* Maseru: Lesotho Government.
Lesotho Government (2012) *National Strategic Development Plan: Growth and Development Strategic Framework.* Washington, DC: International Monetary Fund.
Lesotho Review (2002) Maseru: Wade Publications.
Lesotho Review (2004) Maseru: Wade Publications.
Lesotho Review (2010) Maseru: Wade Publications.
Lesotho Review (2014) Maseru: Wade Publications.
Lukhele, S.E. and Mearns, K.F. (2013) The operational challenges of community-based tourism ventures in Swaziland. *African Journal for Physical, Health Education, Recreation and Dance*, Supplement 2, 199–216.
Maanela, M.T. (2008) Community-based ecotourism for conservation and development in Lesotho: A case of Ha-Kome. MA dissertation, University of Fort Hare.
Manwa, H. (2012) Communities' understanding of tourists and the tourism industry: The Lesotho Highlands water project. *African Journal of Business Management* 6 (22), 6667–6674.
Marie, C.N., Sibelet, N., Dulcire, M., Rafalimaro, M., Danthu, P. and Carrière, S.M. (2009) Taking into account local practices and indigenous knowledge in an emergency conservation context in Madagascar. *Biodiversity and Conservation* 18 (10), 2759–2777.
Mashinini, V. (2003) Tourism policies and strategies in Lesotho. *Africa Insight* 33 (1–2), 87–92.
Mearns, K.F. (2011) Using sustainable tourism indicators to measure the sustainability of a community-based ecotourism venture: Malealea Lodge & Pony Trek Centre, Lesotho. *Tourism Review International* 15 (1–2), 135–147.
Mearns, M.A. (2007) The Basotho cultural village: Cultural tourism enterprise or custodian of indigenous knowledge systems? *Indilinga – African Journal of Indigenous knowledge Systems* 6 (1), 37–50.
Mofokeng, M.J. and Ge, J. (2009) Conservation of endangered species: Aloe polyphylla in Lesotho. *Environmental Research Journal* 3 (2), 68–75.
Mokuku, T. (2012) Lehae-La-Rona: Epistemological interrogations to broaden our conception of environment and sustainability. *Canadian Journal of Environmental Education (CJEE)* 17, 159–172.
Ndlovu, F. (2004) Land reform and indigenous knowledge: A missing link in the fast track land reform programme in Zimbabwe. *Indilinga – African Journal of Indigenous Knowledge Systems* 3 (2), 147–156.
Nedelea, A. and Okech, R. (2008) Developing rural tourism in South Africa. *Bulletin of University of Agricultural Sciences and Veterinary Medicine Cluj–Napoca. Horticulture* 65 (2), 256–261.
Nel, P. (2006) Indigenous knowledge systems, local community and community in the making. *Indilinga – African Journal of Indigenous Knowledge Systems* 5 (2), 99–107.
Pheto-Moeti, M.B. (2005) An assessment of seshoeshoe dress as a cultural identity for Basotho women in Lesotho. MSc dissertation, University of the Free State.
Popova, U. (2014) Conservation, traditional knowledge, and indigenous peoples. *American Behavioral Scientist* 58 (1), 197–214.
Prasad, G. (1998) Scientific knowledge, wild plants and survival. *Agenda: Empowering Women for Gender Equity* 14 (38), 81–84.

Raymond, C.M., Fazey, I., Reed, M.S., Stringer, L.C., Robinson, G.M. and Evely, A.C. (2010) Integrating local and scientific knowledge for environmental management. *Journal of Environmental Management* 91 (8), 1766–1777.

Richards, G. (2007) Introduction: Global trends in cultural tourism. In G. Richards (ed.) *Cultural Tourism: Global and Local Perspectives* (pp. 1–24). New York: The Haworth Hospitality Press.

Saarinen, J. (2006) Traditions of sustainability in tourism studies. *Annals of Tourism Research* 33 (4), 1121–1140.

Saarinen, J. (2009) Sustainable tourism: Perspectives to sustainability in tourism. In J. Saarinen, F. Becker, H. Manwa and D. Wilson (eds) *Sustainable Tourism in Southern Africa: Local Communities and Natural Resources in Transition* (pp. 77–90). Bristol: Channel View Publications.

Samal, P.K., Shah, A., Tiwari, S.C. and Agrawal, D.K. (2004) Indigenous healthcare practices and their linkages with bioresource conservation and socio-economic development in Central Himalayan region of India. *Indian Journal of Traditional Knowledge* 3 (1), 12–26.

Schouten, F. (2007) Cultural tourism: Between authenticity and globalisation. In G. Richards (ed.) *Cultural Tourism: Global and Local Perspectives* (pp. 25–37). New York: The Haworth Hospitality Press.

Seema, J. (2012) The significance of Basotho philosophy of development as expressed in their proverbs. *Indilinga – African Journal of Indigenous Knowledge Systems* 11 (1), 128–137.

Shava, S. (2005) Research on indigenous knowledge and its application. A case of wild food plants of Zimbabwe. *Southern African Journal of Environmental Education* 22, 73–86.

Sigala, M. and Leslie, D. (2005) *International Tourism Management: Implications and Cases*. Oxford: Elsevier.

Smith, M.K. (2009) *Issues in Cultural Tourism Studies*. London: Routledge.

Tanga, P.T. and Maliehe, L. (2011) An analysis of community participation in handicraft projects in Lesotho. *Anthropologist* 13 (3), 201–210.

Taylor, B. and de Loë, R.C. (2012) Conceptualizations of local knowledge in collaborative environmental governance. *Geoforum* 43 (6), 1207–1217.

Teffo, L. (2013) Rural communities as sites of knowledge: A case for African epistimologies. *Indilinga – African Journal of Indigenous Systems* 12 (2), 188–202.

Telfer, D.J. and Sharpley, R. (2008) *Tourism and Development in the Developing World*. London: Routledge.

Timothy, D.J. and Boyd, J.W. (2003) *Heritage Tourism*. Harlow: Prentice Hall.

Timothy, D.J. and Nyaupane, G.P. (2009) Introduction: Heritage tourism and the less-developed world. In D.J. Timothy and G.P. Nyaupane (eds) *Cultural Heritage and Tourism in the Developing World: A Regional Perspective* (pp. 3–19). London: Routledge.

Turner, K.L., Berkes, F. and Turner, N.J. (2012) Indigenous perspectives on ecotourism development: A British Columbia case study. *Journal of Enterprising Communities: People and Places in the Global Economy* 6 (3), 213–229.

United Nations Development Programme/UN World Tourism Organisation (UNDP/UNWTO) (2006) *Support to Institutional and Capacity Strengthening of the Tourism Sector: Final Report*. Maseru, Lesotho.

Viljoen, J. and Tlabela, K. (2007) *Rural Tourism Development in South Africa: Trends and Challenges*. Cape Town: HSRC Press.

Walter, P. (2009) Local knowledge and adult learning in environmental adult education: Community-based ecotourism in Southern Thailand. *International Journal of Lifelong Education* 28 (4), 513–532.

Walter, P.G. and Reimer, J.K. (2012) The 'ecotourism curriculum' and visitor learning in community-based ecotourism: Case studies from Thailand and Cambodia. *Asia Pacific Journal of Tourism Research* 17 (5), 551–561.

Willox, A.C., Harper, S.L. and Edge, V.L. (2013) Storytelling in a digital age: Digital storytelling as an emerging narrative method for preserving and promoting indigenous oral wisdom. *Qualitative Research* 13 (2), 127–147.

4 Narrative and Emotions: Interpreting Tourists' Experiences of Cultural Heritage Sites in KwaZulu-Natal

Joram Ndlovu

Introduction

KwaZulu-Natal is known for its profound history with breathtaking scenery, sandy beaches, unique culture and warm climate. Popularly known as the Zulu Kingdom, it boasts two World Heritage Sites, namely the ISimangaliso Wetland Park (previously known as the Greater St Lucia Wetland Park) and the UKhahlamba-Drakensberg Park. With one of the busiest ports in Africa, Durban has become the economic centre and gateway to a mix of attractions situated within the province. KwaZulu-Natal has a diverse culture, including Indian communities, Zulus and other colonial Europeans. Tourism attractions range from museums, battlefields and cultural villages to Zulu leaders. Of interest are Zulu traditional handicrafts and basket weaving. Within the central and northern part of KwaZulu-Natal, the natural beauty and history are unmasked with the battlefields becoming a symbol of bloody wars that took place in the late 1800s between the Boer commandos, the Zulu armies and the British forces.

KwaZulu-Natal as a Cultural Tourist Destination

George (2001) depicts the province of KwaZulu-Natal as a prime tourism destination with a range of tourist attractions. The Drakensberg area, the cultural histories of the Indian, Zulu and Western people, as well as beaches and museums, are cited as outstanding attractions. The uKhahlamba-Drakensberg Park which has an outstanding natural environment was declared a World Heritage Site in 2000 due to its ancient San rock paintings and overhanging caves. It has an outstanding natural beauty with mixed cultural and natural heritage and is very famous for providing visitors an opportunity to experience a memorable, rich heritage. The innumerable history of the battlefields is rediscovered through historic buildings and museums including Talana Museum. The accounts of battles are emotionally intriguing. The traditional indigenous plants are preserved at the Talana Cultural Garden and Nursery, showcasing different herbs and plants used for medicine and cultural rituals. KwaZulu-Natal is promoted as a magical kingdom, multicultural in character, with the imagery of the Zulu monarchy and the Zulu cultural identity featuring prominently (Seymour, 2006). Shakaland and Kwakunje Cultural Villages portray Zulu traditional culture and lifestyle in a unique way with its traditional Zulu huts, beer pots and beaded ring marks at the entrances. The Zulus are one of the best-known African people internationally through movies, dance myths and folklore. Central to the culture and lifestyle is the animal sculptures and horns which give panoramic views of the history of these people. The Zulus had a huge dominion across the history of southern and eastern Africa, the influence of which shaped the political and social structure of its people. Some notable attractions include the graves of the Zulu kings and Zulu leaders, traditional Zulu ceremonies and Zulu cultural villages, including contemporary leaders such as Albert Luthuli, who was an extraordinary person who spent his life in the service of humanity. Luthuli Museum honours a man who contributed immensely to the struggle for freedom with luminaries such as Nelson Mandela and Langalibalele Dube (Marschall, 2012).

The Inanda Heritage Route provides a fascinating spectacle of history. The route, which takes you through the Ohlange Institute, the Phoenix Settlement of Ghandi, Inanda Seminary and Ebuhfeni, the place of Shembe Church, gives a unique glimmer of the historical insights in these significant cultural heritage sites. The Phoenix Settlement is officially positioned by the government and the Phoenix Settlement Trust as a symbol of unity, reconciliation and peace, yet it is a contested and highly politicised site overshadowed by a dark history of racial violence, which is still deeply lodged in local community memory and oral history (Marschall, 2008). Apart from the five historical nodes, Inanda contains a rich body of intangible and 'living' heritage, much of it rooted in traditional Zulu culture, including traditional healing, music, dancing and craft-making, notably beadwork (Marschall, 2012).

Tourists enter into the KwaZulu-Natal scene with a heady sense of 'being there', capturing an essential part of the life, culture of the locality, its traditions and capturing some sense of 'real' meaningful experience in action, a ritual in motion (McCabe & Stokoe, 2004). The emotions fostered make the exploration of culture and heritage resources a dynamic activity, developed through physical experiences, searching and celebrating what is unique and beautiful, represented by people's own values and attributes which are worth preserving as an inheritance for their descendants (Vargas-Hernández, 2012). The above narrative shows that heritage and culture are closely linked and ardently position tourism at the centre of history. Hence any debate on both tangible and intangible assets collides and intersects in the staging, or re-enactment, of historical events (McCabe & Foster, 2006). Due to its distinctiveness, the Global Competitiveness Report (2009) shows that there is an emerging interest in cultural heritage, where foreign tourists would prefer to participate interactively with local cultures and take part in township tours and homestays.

Historical Re-enactment Sites for Tourist Consumption

Re-enactment can be defined as a way of reconstructing the past in the present – a specific war or other event – or it may be more broadly defined by a social movement or period framed by specific years or significant personalities. Therefore, re-enactment is a form of live action and role playing, it spans diverse history-themed genres, from theatrical and living history performances to museum exhibits, television and films. The benefits of historical re-enactments are monumental. These vary from unlocking the potential of culture and heritage to sustainable and responsible tourism development (Marschall, 2012). From a tourism perspective, historical re-enactment does not only bring economic benefits but it can lead to social cohesion. However, for this to be realised, cultural heritage resources need to be systematically aligned with mainstream tourism products. The challenge is how to integrate these products and represent them on South Africa's promotional marketing materials (Nzama et al., 2005). The current literature and interest is vested in other forms of tourism, with culture and heritage tourism becoming an insignificant part in the whole tourism phenomenon, whose value and impact have not been fully investigated. As a result, misrepresentation of culture and heritage through uninformed interpretations at different heritage sites tends to compromise the authenticity and integrity of cultural and heritage products. The disparities between development requirements and conservation may further widen the gap. To fully comprehend what is happening in cultural and heritage tourism, there is a need to develop an integrated framework that spells out revenue streams, linking these to culture and heritage tourism.

Tourists' Motivation to Travel to KwaZulu-Natal

Cultural heritage tourism (CHT) is viewed as travel concerned with experiencing cultural environments, including landscapes, the visual and performing arts, and special lifestyles, values, traditions and events (Baxter & Purcell, 2007; Duxbury & Jeannotte, 2010). It is important to stress that CHT involves not only tangible or visible heritage such as sites, colours, materials and settlement patterns, but also intangible heritage such as societal structures, traditions, values and religion (Endresen, 1999; McCool & Moisey, 2008). To be precise, heritage tourism refers to tourists travelling to places of traditional, historical and cultural significance with the aim of learning – paying respect for recreational purposes (Nzama *et al.*, 2005). Travel behaviour can therefore be defined as a way tourists behave based on their attitudes towards a certain product and their response to it and use of it (George, 2001; Van Vuuren & Slabbert, 2011). Motivational theories are also discussed by Crompton (1979), who identified two layers of socio-psychological motivation, while Botha *et al.* (1999) pursued personal motivations (push factors), destination attributes (pull factors) and situational inhibitors. For instance, most tourists visiting culture and heritage sites in KwaZulu-Natal are driven by motives such as the need to slow down and get away from their daily routine. This is evident in Dellaert *et al.*'s (1998) research which developed a constraint-based conceptual framework that sought to describe tourists' sequential choices of travel components. The feeling of escape from monotony at home brings with it a sigh of relief. Since motivational factors are a significant source of disequilibrium, they can only be corrected through a tourism experience (Dann, 1977). Consequently, some of the push factors driving tourists to visit KwaZulu-Natal include nostalgia, rurality and emotion while others take the opportunity that they have for visiting their friends and relatives to visit cultural heritage sites.

Some tourists travel to a destination like Durban for meetings, events or conferences, as a result they end up visiting cultural heritage sites. Tourists can be classified according to various categories including purpose of trip, type of attraction visited, type of accommodation preferred (McKercher, 2002; Mohammad et al., 2010; Richards, 1996; Stebbins, 1996). Tourists, for instance, typically travel for pleasure, and therefore view their world through what might be considered a hedonistic cultural lens which defines the locality in relation to the degree with which it fulfils their leisure needs and matches their expectations (Jones, 2008; Jönsson & Devonish, 2008). There is a general agreement on the push and pull factors, although some researchers (Cohen, 2004; Urry, 1995) have discussed travel motivation in relation to the tourist gaze. A tourist gaze expresses the dynamics associated with the construction of a tourist experience, the complexity of the social organisation of tourism, and the systematic nature of these processes. It allows us to articulate what separates tourist experience from everyday living and illuminates ways in which production and consumption of tourist goods and services have wide implications for social relations. In this world of spectacle there is nothing original, no real meaning, everything is a copy, or text upon a text. It is a depthless world of networks of information and communication in which information has no final purpose in meaning (Urry, 1988). The tourist gaze suggests that tourist experience involves a particular way of seeing. Images and myths about what to see tend to be distinctive, striking, unusual and extraordinary. Such visual and narrative depictions of tourist destinations are strategically promoted by the marketing industry to contrast with people's daily routine and work schedules at home.

The niche area of cultural and heritage tourism, comprising township tours, cultural villages, battlefield tours and the like, is believed to hold particular promise for racial transformation of the tourism sector and the empowerment of previously marginalised communities, as this type of tourism is structured around existing, often community-based resources (Marschall, 2012). As such, Hunt (2004) identifies culture and heritage sites as the most suitable leisure pursuit where travellers get an opportunity to engage creatively to become participants. As a result, tourists tend to become emotionally attached to and identify with battlefields and museums, Zulu shrines, sculptures and curios made locally. The experiences, feelings and sentiments that are cited most often about KwaZulu-Natal are that the destination is intriguing, educational and authentic. In essence, cultural tourists contemplate the ways in which a series of contrasts between the present and the past, between a town and a commune or the stereotypes of everyday life and the exotic are expressed (Petroman et al., 2013: 385). However, the behavioural characteristics and passions of cultural tourists are different from other visitors on most measurements and this has an implication for the development and promotion of this niche market segment.

Recently, the post-tourist concept has come under scrutiny. With the advent of globalisation, cultural capital has realised that tourist activities are staged and yet still revel in the inauthenticity and kitsch offered by the performances (MacCannell, 2001). Globalisation as a concept refers to 'a process of growing interconnectedness between people as a result of the decreased effects of distances and political boundaries and is marked by a reduced State role and an increase in the role of non-state actors. There are two approaches that will aid in our understanding: the convergent approach and the divergent approach' (O'Neill, 2002: 519). From a convergent approach, the results of globalisation show that countries are continually outward looking, developing strategies that focus on local development whilst taking on investment opportunities, community participation, employment creation and cultural preservation. Consequently, some tourists visiting cultural heritage sites in KwaZulu-Natal tend to have a vested interest in history; the need to refresh old memories, and learn more about self and others becomes a driving force behind travel. Some reasons for travel include visiting issues of reflection and meditation, learning about culture and tradition, experiencing something new and even satisfying curiosity (Cohen, 1972). The literature reveals that, rather than having the ability or even the desire to go anywhere and do anything, the post-tourist knows all and sees all. Based on stories, tales or narratives he/she moves from one tourist role/performance to another (Harvey & Lorenzen, 2006). However, to date there is little empirical evidence on the push and pull factors for visiting re-enactment sites in KwaZulu-Natal. However, from a divergent view, whilst globalisation can promote locally owned enterprises, because of free trade and weak national capital these cannot compete with strong foreign companies which often undermine the sustainability of smaller, locally owned tourism ventures in developing countries (O'Neill, 2002: 520).

Cultural and Heritage Tourism: Prospects, Opportunities and Challenges

The World Tourism Organisation (WTO) predicts that cultural tourism will be one of the five key tourism markets in the future, and notes that growth in this area will present an increasing challenge in terms of managing visitor flows to cultural sites (Endresen, 1999). Cultural tourism has long existed, but recent demographic, social and cultural changes in the main source countries have led to an increasing number of new niche markets in different destinations, thereby making culture and heritage crucial to people's identity, self-respect and dignity (Endresen, 1999). Simultaneously, these cultural landscapes are being staged and commoditized as past experiences for leisure and recreational activities (Correia, Oom Do Valle & Moço, 2007) in South Africa. The 'living history' has become an important educational tool, and also an important part of contemporary leisure life for participants and spectators as well as educators and historians (McCabe & Foster, 2006). Cultural tourism is defined as the movement of persons for essentially cultural motivations, which include study tours, performing arts, cultural tours, travel to festivals, visits to historic sites and monuments, folklore and pilgrimages (WTO, 1985). Although sun, sand and surf holidays are not expected to disappear, they have declined in relative importance as more and more visitors seek challenging, educational and/or relatively unique experiences (Endresen, 1999). Cultural tourism is a tourism product by itself and can contribute highly to regional economic development (Vargas-Hernández, 2012). Therefore, developing a sustainable cultural tourism opportunity has to strive for balance of the benefits for all communities involved – the tourism community, the cultural sector providers and the host community that forms the background to the cultural product (Vargas-Hernández, 2012). As such, cultural tourism has become the latest buzzword in cultural heritage tourism marketing.

The tourist gaze presents a complex and dynamic social representation associated with tourist experiences (Cohen, E. & Cohen, S.A., 2012). The reconstructed experiences allow tourists to understand the lived experiences of certain people with varied social relations implications. As Urry (1988: 39) explains, 'this world of spectacle is one in which there is nothing which is original, no real meaning, everything is a copy, or text upon a text. It is a depthless world of networks of information and communication in which information has no end purpose in meaning'. Not only does the tourist gaze suggest seeing things differently, what is perceived, myths and imageries are extraordinarily different (Urry, 1990). Destinations are depicted and promoted by contrasting people's lifestyles with daily work routines. Hence, the niche area of cultural and heritage tourism, comprising township tours, cultural villages, battlefield tours and the like, is believed to hold particular promise for the racial transformation of the tourism sector and the empowerment of previously marginalised communities, as this type of tourism is structured around existing, often community-based resources (Marschall, 2012; Nzama et al., 2005; WTO, 1985). Therefore, the growth of this market has influenced the development of historical re-enactment sites. Hunt (2004) identifies historical re-enactment as a leisure pursuit and defines it as a form of tourism where travellers get an opportunity to engage creatively to become participants. Studies have shown that culture and cultural heritage are crucial to people's identity, self-respect and dignity (Endresen, 1999). Therefore, cultural tourism is a tourism product by itself and can contribute highly to regional economic development (Vargas-Hernández, 2012). As such, it has become a buzzword for attracting tourists to visit cultural heritage sites. The sustainability of this form of tourism lies in its ability to strike a balance between community benefits, the tourism industry and cultural heritage resource providers. 'The potential benefits of World Heritage extend far beyond the sites which have been listed, since these areas can play a leadership role in setting standards for protected areas as a whole, can bring resources for training which will be of wider application, and can be "flagships" in terms of raising public awareness of conservation issues' (Endresen, 1999).

Although the economic benefits of heritage and cultural tourism tend to be disproportionate and uneven in a particular locality, this segment of tourism has the potential to generate a significant number of decent jobs which are vital for combatting the high unemployment rate in a country like South Africa. However, throughout South Africa, museums and heritage sites struggle to attract visitors from previously disadvantaged communities, despite the frequent absence of an entrance fee (Marschall, 2012). Baker and Crompton (2000) argue that, although a substantial literature has been written in this area, relatively little discussion has made a distinction between the constructs of quality of performance and the level of tourist satisfaction, nor has there been any assessment of their relative impact on subsequent behaviour.

Considering the past 30 years, tourism has become the world's fastest-growing economic sector, and South Africa is ranked amongst the world's 25 top tourist destinations (Marschall, 2008). With the flourishing of cultural and heritage tourism in South Africa, cultural heritage (however that might be defined) has been identified as a precious resource, for preservation, celebration and commercialisation. Currently, the traditional tourism markets for South Africa are based on international tourists, whose main interests are usually inclined towards mainstream tourism products. Most promotional strategies are targeted to overseas visitors, most of whom pay less attention to those events considered to be less mainstream. To ensure full integration of this product, destination marketers need to develop promotional materials that capture the cultural essence of a place and make it more accessible to all. By developing cultural and heritage programmes that define people's insights of life, a destination can position itself competitively and maximise opportunities of developing this sector progressively.

Conclusion

The tourist gaze presents a complex and dynamic social representation associated with tourist experiences. The reconstructed narratives in a destination allow tourists to understand the lived experiences of certain people with varied social relations implications. Cultural tourism as a leisure pursuit has become a buzzword and gives tourists an opportunity to engage creatively to become participants in historical events, which is crucial to people's identity, self-respect and dignity. Therefore, travel motives form an integral part of travel behaviour and have been widely researched and applied in tourism marketing strategies but with little applicability to cultural heritage sites.

The chapter concludes that KwaZulu-Natal is becoming a favourable cultural tourism destination, with participants expressing feelings of satisfaction with the essential amenities. The stories and tales about Zulu culture and heritage provide feelings of nervousness and excitement. Therefore, educating local communities to become conduits of knowledge about local heritage can provide emotional appeal and fascinating feelings that are potentially provocative, humorous, passionate, exciting and insightful for tourists. Consequently, this can substantially enrich the quality of visitors' experiences.

The managerial implications are that the continued growth of culture and heritage as a niche tourism product has an implication for sustainable tourism development. Therefore, destination management organisations should play an important role in product development and cultural heritage tourism promotion by providing opportunities to locals to interact with tourists. Although complex, tourism products and services can provide compelling stories with an emotional pull to the destination. Since cultural and heritage tourist typologies are different from other forms of tourism, a coordinated approach to developing common, innovative and creative marketing strategies is essential in providing unique historical experiences.

References

Baker, D.A. and Crompton, J.L. (2000) Quality, satisfaction and behavioral intentions. *Annals of Tourism Research* 27 (3), 785–804.
Baxter, H.K. and Purcell, M. (2007) Community sustainability planning. See www.naturalstep.ca (accessed 31 October 2014).
Botha, C., Crompton, J.L. and Kim, S. (1999) Developing a revised competitive position for Sun/Lost City, South Africa. *Journal of Travel Research* 37, 341–352.
Cohen, E. (1972) Toward a sociology of international tourism. *Social Research* 39 (1), 164–182.

Cohen, E. (2004) *Contemporary Tourism: Diversity and Change*. New York: Elsevier.
Cohen, E. and Cohen, S.A. (2012) Current sociological theories and issues in tourism. *Annals of Tourism Research* 39 (4), 2177–2202.
Correia, A., Oom Do Valle, P. and Moço, C. (2007) Why people travel to exotic places. *International Journal of Culture, Tourism and Hospitality* 1 (1), 45–61.
Crompton, J.L. (1979) Motivation for pleasure vacation. *Annals of Tourism Research* 6, 408–424.
Dann, G.M.S. (1977) Anomie, ego-enhancement and tourism. *Annals of Tourism Research* 4 (4), 184–189.
Dellaert, B.G.C., Ettema, D.F. and Lindh, C. (1998) Multi-faceted tourist travel decisions: A constraint-based conceptual framework to describe tourists' sequential choices of travel components. *Tourism Management* 19 (4), 313–320.
Duxbury, N. and Jeannotte, S.M. (2010) Culture, sustainability, and communities: Exploring the myths. Centre de Estudes Socias (CES), Facuidade de Economia, Universidade de Cooimbra, Oficina do CES no. 353. Setembro de 2010. See www.ces.uc.pt/myces/UserFiles/livros/614_CES%20Oficina_353.pdf (accessed 15 March 2015).
Endresen, K. (1999) Sustainable tourism and cultural heritage: A review of development assistance and its potential to promote sustainability. See www.nwhf.no/files/File/culture_fulltext.pdf (accessed 29 March 2014).
George, R. (2001) *Marketing South African Tourism and Hospitality*. Cape Town: Oxford University Press.
Global Competitiveness Report (2009) *Global Competitiveness Report 2009–2010*. Geneva, Switzerland: World Economic Forum.
Harvey, D.C. and Lorenzen, J. (2006) Signifying practices and the co-tourist. *Tourismo: An International Multidisciplinary Journal of Tourism* 1 (1), 11–28.
Hunt, S. (2004) Acting the part: 'living history' as a serious leisure pursuit. *Leisure Studies* 23 (4), 387–403.
Jones, D. (2008) Beyond a tourist gaze? Cultural learning on a tourist semester abroad programme in London. *Journal of Research in International Education* 7 (1), 21–35.
Jönsson, C. and Devonish, D. (2008) Does nationality, gender, and age affect travel motivation? A case of visitors to the Caribbean island of Barbados. *Journal of Travel & Tourism Marketing* 25 (3–4), 398–408.
Marschall, S. (2008) An inspiring narrative with a shadow: Tangible and intangible heritage at the Phoenix Settlement of Mahatma Gandhi. *Southern African Humanities* 20, 353–374.
Marschall, S. (2012) Sustainable heritage tourism: The Inanda Heritage Route and the 2010 FIFA World Cup. *Journal of Sustainable Tourism* 20 (5), 721–736.
McCabe, S. and Foster, C. (2006) The role and function of narrative in tourist interaction. *Journal of Tourism and Cultural Change* 4 (3), 194–215.
McCabe, S. and Stokoe, E. (2004) Place and identity in tourist accounts. *Annals of Tourism Research* 31 (3), 601–622.
McCool, F.S. and Moisey, N.R. (2008) *Tourism, Recreation and Sustainability: Linking Culture and the Environment*. Oxfordshire: CABI Publishing.
McKercher, B. (2002) Towards a classification of cultural tourists. *International Journal of Tourist Research* 4, 29–38.
Mohammad, A.B., Mohammad, A. and Mat Som, P.A. (2010) An analysis of push and pull travel motivations of foreign tourists to Jordan. *International Journal of Business and Management* 5 (12), 41–50.

Nzama, A.T., Magi, L.M. and Ngcobo, N.R. (2005) Workbook-I Tourism Workbook for Educators: 2004 Curriculum Statement. Unpublished Tourism Workshop Educational Materials. Centre for Recreation and Tourism, University of Zimbabwe and Tourism KwaZulu-Natal.

O'Neill, A.C. (2002) What globalization means for ecotourism: Managing globalization's impacts on ecotourism in developing countries. *Indiana Journal of Global Legal Studies* 9 (2), Article 6. See http://www.repository.law.indiana.edu/ijgls/vol9/iss2/6 (accessed 19 August 2015).

Petroman, I., Petroman, C., Marin, D., Ciolac, R. Văduva, L. and Pandur, I. (2013) Types of cultural tourism. *Animal Science and Biotechnologies* 46 (1), 385–388.

Richards, G. (1996) Production and consumption of European cultural tourism. *Annals of Tourism Research* 23 (2), 261–283.

Seymour, J. (2006) The Marketing and Management of Tourism KwaZulu Natal. An interview held with James Seymour the Marketing and Research Manager TKZN. Held Durban in May 2006. In K.M. Xulu (ed.) Indigenous Culture, Heritage and Tourism: An Analysis of the Official Tourism Policy and its Implementation in the Province of Kwazulu-Natal. Unpublished PhD Thesis, University of Zululand.

Stebbins, R.A. (1996) Cultural tourism as serious leisure. *Annals of Tourism Research* 23 (4), 948–950.

Urry, J. (1988) Cultural change in contemporary holiday making. *Theory, Culture and Society* 5, 35–55.

Urry, J. (1990) *The Tourist Gaze*. London: Sage Publications.

Urry, J. (1995) *Consuming Places*. New York: Routledge.

Van Vuuren, C. and Slabbert, E. (2011) Travel behaviour of tourists to a South African holiday resort African. *Journal for Physical, Health Education, Recreation and Dance* 17 (4), 694–707.

Vargas-Hernández, G.J. (2012) A normative model for sustainable cultural and heritage tourism in regional development of Southern Jalisco. *Innovative Journal of Business and Management* 1 (1), 5–15.

World Tourism Organization (WTO) (1985) *The State's Role in Protecting and Promoting Culture as a Factor of Tourism Development and the Proper Use and Exploitation of the National Culture of Heritage of Sites and Monuments for Tourism*. Madrid: WTO.

5 Cultural Heritage Tourism Development in Post-Apartheid South Africa: Critical Issues and Challenges

Gareth Butler and Milena Ivanovic

Introduction

The aim of this chapter is to present an overview of current strategic and experiential challenges arising from the development of cultural heritage tourism in South Africa. Despite the recent development of a number of new tourism strategies (National Heritage and Cultural Tourism Strategy, 2011, 2012; National Tourism Service Strategy, 2011; Rural Tourism Strategy, 2011) that are designed to directly or indirectly facilitate further growth of cultural heritage tourism in South Africa, numerous problems still persist. The chapter begins with an overview of the current state of cultural heritage tourism in South Africa. This is then followed by a discussion on the issues arising from inadequate strategic planning and an Afrocentric developmental focus. The discussion is supported by data from South African Tourism's annual reports (SAT, 2007–2013), the South African Global Competitiveness study (2004/2005), and selected case studies.

Although the cultural heritage sector continues to attract visitors and generate economic benefits, its overall contribution to South African tourism remains unknown because of poor national statistical record keeping, making any benchmarking of the current National Culture and Heritage Tourism Strategy (NDT, 2012a) difficult. A further concern is that South African cultural heritage tourism products have been recently described as being 'substandard' in numerous government reports relating to visitor experience and management. This chapter concludes by providing a series of recommendations that may stimulate enhanced tourist experiences, and more broadly, the future success of the cultural heritage tourism sector in South Africa.

The Importance of Cultural Heritage Development

Cultural heritage tourism is one of the most prevalent types of tourism in the world (van der Ark & Richards, 2006: 1408). Recent estimates suggest that this particular form of travel accounts for approximately 40% of global tourism (ATLAS, 2009: 98; OECD, 2009: 21). The significant growth of cultural heritage tourism has been driven by a shift in production and consumption from Fordist economies of scale associated with conventional mass tourism (CMT) to Post-Fordist economies of scope, also known as the experience economy (Pine & Gilmore, 1999). The new experience economy gave rise to individually orientated production in the form of niche tourism or special interest tourism(s) (SITs). These modes of tourism reflect 'consumption characterized by deeper preferences for small-scale and high class products' (Saarinen et al., 2009: 6). The authentic experiential value of a destination's cultural heritage has emerged as a key priority, shifting developmental and competitive focus from provision of services to the facilitation of authentic and differentiated tourist experiences. Longitudinal research findings confirm that cultural heritage attractions dominate tourists' consumption patterns with museums rated 'the most important' tourist attractions by 65% of visitors worldwide, followed by historical sites (52%) and monuments (48%) (ATLAS, 2007: 5). Further evidence of the 'omnipotence' (Richards, 2007: 1) of the cultural heritage sector lies in the economic benefits it can generate at destinations (Garrod & Fyall, 2000: 683). Many countries have already successfully employed cultural heritage as both a vehicle for urban renewal and economic growth, and as a tool of regional economic development, both in terms of generated revenues and employment (Massyn & Koch, 2004; Ndlovu & Rogerson, 2004).

The development of the cultural heritage tourism sector in South Africa has the potential to yield numerous benefits too, including those that venture beyond economic growth. Cultural heritage may be used as an opportunity to positively influence how the international community views a particular nation (Alvarez & Korzay, 2011; Hughes & Allen, 2005; Light, 2001). Indeed, this approach does not differ too greatly from South Africa's hosting of the 2010 FIFA World Cup, which, to some extent, led to a reassessment of global views on the country as a tourism destination (Lepp & Gibson, 2011; Woodward & Goldblatt, 2011). Cultural heritage sites have the power to act as 'tangible canvasses for intangible constructs' to be developed (Butler *et al.*, 2014: 204) and enable positive national narratives to be interwoven through material and socio-psychological experiences (Garrod & Fyall, 2000, 2001; McIntosh & Prentice, 1999; Park, 2010, 2011; Poria & Ashworth, 2009). Similarly, cultural heritage tourism has also been identified as a vehicle for 'nation-building' (Light, 2007: 747) and as a potential mechanism for feelings of national inclusivity to be formed (Palmer, 1999; Pretes, 2003). Thus, it has been argued that developing collective identities is often 'an official goal of countries comprised of many different immigrant cultures' (Pretes, 2003: 125), and a similar rhetoric is frequently found in South African policies and strategies that aim to foster cultural heritage tourism growth. Selecting which cultural heritage sites should act as transmitters for a collective national identity in the 'new' South Africa, remains a crucial point of debate and one which requires further discussion. However, a range of other challenges also persist and South Africa must negate these if cultural heritage is to be successfully developed. We continue by providing a brief appraisal of the current state of heritage tourism in South Africa.

The Current State of Cultural Heritage Tourism in South Africa

According to SAT's 2012 Annual Tourism Report (SAT, 2013), of the 9.19 million international arrivals recorded, only 18.7% (approximately 1.72 million) of tourists arrived primarily to travel for the purpose of a holiday. Instead, visiting friends and relatives (VFR) remained the dominant market for international tourists, accounting for 27.7% of all visits (see Table 5.1).

Table 5.1 Purpose of visit to South Africa, 2012 (international tourists)

Purpose of visit	*Percentage (%)*
Visiting friends and relatives	27.7
Holiday	18.7
Business	18.4
Shopping (Personal)	15.6
Shopping (Business)	10.3
Medical	3.0
Religion	0.7
Other	5.7

Source: SAT (2012).

However, 43.1% of international tourists visiting South Africa primarily for the purpose of a holiday participated in trips to cultural, historical or heritage sites during their visits. For tourists primarily engaged in South Africa's largest sector, VFR, 20.3% of this visitor type stated that they too had visited cultural, historical or heritage sites during their journeys.

Although it is clear that visiting cultural heritage attractions may not be the dominant motivation factor for visiting the country, South Africa's cultural heritage remains an important facet of the national tourism economy (Briedenhann & Wickens, 2004). Using conservative estimates based on SAT statistics (2012), it is reasonable to estimate that at least 1.5 million international tourists will typically engage in cultural heritage tourism during their stay. Moreover, as Binns and Nel (2002) have observed, government and private entrepreneurs have been quick to identify the considerable potential of cultural heritage in South Africa and have sought to develop it further. Indeed, the nation's cultural heritage has been identified as an excellent opportunity and a 'rapid growth area' for job creation and empowerment to occur (Gössling, 2000; Phaswana-Mafuya & Haydam, 2005). However, in order to realise this potential, it is imperative that the South African government addresses a number of key debates. These prominent themes will now be discussed in further detail.

Box 5.1 Industrial heritage tourism – the 'Big Hole', Kimberley, South Africa

Clinton David van der Merwe

Heritage tourism is a growing segment of cultural tourism within the global tourism economy. Increasingly, heritage tourism is being identified as one of the major growth markets in global tourism (Tlabela & Munthree, 2012) and some argue that heritage tourism in developing countries has the potential to valuably contribute to local economic development (LED) (van der Merwe & Rogerson, 2013). Heritage tourism includes both the tangible and intangible elements of the past and is defined by many scholars from both a supply as well as a demand perspective (see Poria *et al.*, 2001; Timothy & Nyaupane, 2009).

There are various segments of heritage tourism, with industrial heritage tourism being one of the most interesting, which form a distinctive, but under-researched, subset of the wider field of heritage tourism. Industrial heritage tourism is defined as 'the development of touristic

(continued)

Box 5.1 Industrial heritage tourism – the 'Big Hole', Kimberley, South Africa (*continued*)

activities and industries on a man-made (sic) site, buildings and landscapes that originated with industrial processes of earlier periods' (Edwards & Llurdes i Coit, 1996: 342). While others see industrial heritage as that which 'refers to housing, industrial settlements, industrial landscapes, products and processes and documentation of the industrial society' (Xie, 2005: 1321). Most scholars concur that the global rise of industrial heritage tourism is associated with the deindustrialisation and the growth and development in the leisure industry since the 1970s, which encouraged the 'heritagization' of formerly industrial places (Walsh, 1992). Many cities in developed countries try to use industrial heritage to develop their local economies, as they endeavour to rebuild themselves and sustainably utilise their local resources and heritage assets (Law, 1993). Industrial heritage tourist activities contribute to preserving 'a region's identity and to stimulate the formation of local service activities and employment' (Hospers, 2010: 398).

Beyond the remodelling of factories or docklands, much industrial heritage tourism relates to the use of former mining operations. Overall, the international trend is towards 'the conversion of mining valued for industrial purposes to mining valued for its heritage and tourism aspects' (Conlin & Jolliffe, 2011: 3). With its long mining history, South Africa has considerable potential for the growth of industrial heritage tourism; the 'Big Hole' at Kimberley is one of the top tourist attractions in the Northern Cape Province (Plate 5.1). The Kimberley mine, fabled as 'the Diamond Fields', has a long history of diamond discovery; it operated from the 31 March 1871–4 August 1914. Since that time the mine area has become a tourist attraction (van der Merwe & Rogerson, 2013).

In 1968 the Kimberley Mine Museum was established by De Beers Consolidated Mines as a social history museum and in 2006 re-developed as The 'Big Hole' Project (Brown, 2006). This was a major initiative towards making the industrial heritage of Kimberley a favourable tourist attraction. Local tourism marketing proclaims it as: 'The frenetic activity, the extraordinary web of pulley cables leading to a six-storey staging platform and the sight of up to 30,000 miners working 3600 claims over 17ha have faded into the sepia memories of photographic archives. Yet, somehow, memories of the Kimberley tent-town's days linger. Many old buildings, museums and one of South Africa's most important art

Plate 5.1 The Big Hole, 2013
Source: Photo by Clinton David van der Merwe.

galleries lend an historic ambience to the city that thrust its way to prominence during the diamond rush. A reconstruction of the original "rush town" stands alongside the incredible Big Hole, the largest hand-dug excavation in the world, offering visitors insight into the lives of those who lived and worked through the dreams and nightmares of a vibrant history we take for granted' (Northern Cape Tourism Authority, 2013).

The 'Big Hole' is owned by De Beers Consolidated Mines, which set aside R60 million (€6 million) to transform 'the Kimberley "Big Hole"' from a worn out-mine into a tourist destination. De Beers Consolidated Mines stated 'We want this to be a major attraction'. Furthermore, it was made clear that: 'Now that the mines are closing, a number of people in Kimberley will lose their jobs. We would like to soften that blow by creating an alternative industry' (Bates, 2004). Significantly, many of the museum employees are ex-mine personnel or contractors. As Brown (2006: 4) avers, the 'current thinking behind the new attraction is focused largely on its commercial sustainability and corporate image and hence on its ability to attract and entertain an audience based on tourism'.

(continued)

Box 5.1 Industrial heritage tourism – the 'Big Hole', Kimberley, South Africa (*continued*)

The Kimberley 'Big Hole' was officially opened on 6 November 2006 (The Big Hole, 2013). The visitor experience includes, *inter alia*, a short film which introduces the story of diamonds at Kimberley in a state-of-the-art 65-seater auditorium; an underground mine experience; a real diamond display; as well as being able to see the 'Big Hole' from a viewing platform; a walk through the pulsator building and the exhibition centre; and a nostalgic stroll through the Old Mining Town.

References

Bates, R.B. (2004) One of the world's first diamond mines may become a tourist attraction. Blog. See www.jckonline.com (accessed 24 January 2007).

Brown, M. (2006) Re-envisioning the Kimberley Mine Museum: De Beers' Big Hole Project. Unpublished MA dissertation, University of the Witwatersrand, Johannesburg.

Conlin, M.V. and Jolliffe, L. (eds) (2011) *Mining Heritage and Tourism*. Oxford: Routledge.

Edwards, J.A. and Llurdes i Coit, J.C. (1996) Mines and quarries: Industrial heritage tourism. *Annals of Tourism Research* 23 (2), 341–363.

Hospers, G.J. (2010) Industrial heritage tourism and regional restructuring in the European Union. *European Planning Studies* 10 (3), 397–404.

Law, C.M. (1993) *Urban Tourism: Attracting Visitors to Large Cities*. London: Mansell.

Northern Cape Tourism Authority (2013) Diamond Fields Visitors Guide. See http://experiencenortherncape.com/visitor/explore-the-northern-cape/regions/diamond-fields (accessed 1 July 2013).

Poria, Y., Butler, R. and Airey, D. (2001) Clarifying heritage tourism. *Annals of Tourism Research* 28 (4), 1047–1049.

The Big Hole (2013) *The Big Hole Kimberley – Diamonds & Destiny*. Pamphlet: Swift Print.

Timothy, D.J. and Nyaupane, G.P. (eds) (2009) *Cultural Heritage and Tourism in the Developing World: A Regional Perspective*. London: Routledge.

Tlabela, K. and Munthree, C. (2012) An investigation into tourists' satisfaction with culture and heritage tourism in South Africa – An exploratory study. *International Journal of Culture and Tourism Research* 5 (2), 1–9.

van der Merwe, C.D. and Rogerson, C.M. (2013) Industrial heritage tourism at the 'Big Hole', Kimberley, South Africa. *African Journal of Physical, Health Education, Recreation and Dance* 19 (Supplement 2), 155–171.

Walsh, K. (1992) *The Representation of the Past. Museums and Heritage in the Post-Modern World*. London: Routledge.

Xie, P.F. (2005) Developing industrial heritage tourism: A case study of the proposed Jeep museum in Toledo, Ohio. *Tourism Management* 27, 1321–1330.

South African Cultural Heritage: Opportunities and Challenges

One of the three objectives of the 2011 National Tourism Service Strategy (NTSS) involved the need to create 'an enhanced visitor experience'. One of its related targets included the aim 'to deliver world-class visitor experience' (NDT, 2011c: 19), incorporating the intention to deliver 'tourist experiences that equal or surpass the expectations of foreign and domestic tourists alike' (NDT, 2011c: 19). However, despite these bold aims, South African cultural heritage products have often performed poorly according to government research on visitor experience. Indeed, the reports reveal that according to international tourists, South Africa's cultural products 'lack[ed] authenticity' (49% of visitors agreed with this view), were 'not sophisticated enough' (46% agreed), and failed to provide 'empathy with the cultural product' (42% agreed) (DEAT & DTI, 2004: 367). Worryingly, further reports have noted a marked decline in the number of international tourists that rated South Africa's cultural heritage as the 'best experience' during their stays (SAT, 2008, 2009, 2010, 2011, 2012).

In 2012, an NDT report was developed to record the experiences, perceptions and expectations of consumers of cultural tourism products. The report revealed that many international tourists visiting South Africa were 'not specifically culturally motivated' tourists (NDT, 2012b: 49). Although the research of Ivanovic and Saayman (2013) echoed these observations to some degree, it was still noted that at least 20% (approximately 500,000) of all international tourists to South Africa considered themselves to be highly motivated cultural tourists. Despite these observations, the NDT suggested that cultural tourism products should instead be developed to meet the demands of 'accidental' or low-motivated cultural tourists. This approach contradicts the views of McKercher (2002) who argued that it is an often incorrect approach to classify some visitors as being 'accidental cultural tourists', for they will still harbour motivations to engage with cultural heritage products. Moreover, to assume that visitors' experiential demands are low may evidently lead to poor product development. Indeed, developing cultural tourism products for visitors that are assumed to be 'casual consumers', perhaps explains to some extent the poor experiential values recorded. We continue by exploring a range of problematic factors which may also further illustrate the unfavourable experiences of international tourists.

Finding a representative South African cultural heritage

Despite now being in the second decade of transition since the end of the apartheid era, questions of developing a representative *South African* identity remain at the forefront of political debate (Bredekamp, 2006; Coombes, 2003; van der Waal & Robins, 2011). South Africa is represented by 11 different official languages and, although four broad racial groups are predominantly used in official documentation (African, White, Coloured and Asian), these groups are also complex in nature. For example, Black South Africans constitute the Nguni, the Sotho-Tswana, the Tsonga and the Venda, which may still be subdivided further, while White South Africans are typically considered an amalgamation of Dutch, French Huguenot and British settlers that arrived from the 16th century onwards. While the broad spectrum of ethno-racial groups poses an obvious barrier to the construction of a singular national identity, South Africa's history of apartheid undoubtedly compounds matters further. In response to these challenges, a White Paper on Arts, Culture and Heritage was developed in June, 1996 under the Mandela administration (DEAT, 1996). The White Paper sought to redefine and re-imagine South Africa's cultural heritage, including a radical shift in focus to incorporate non-White cultural heritage that had been significantly diluted during apartheid (Bredekamp, 2006). Although the newly established democratic South Africa inherited a cultural heritage portfolio that consisted of a wide spectrum of museums, archaeological sites, diverse music and arts, and rituals, many of these had 'not [been] used to the maximum benefit of society at large' (Phaswana-Mafuya and Haydam, 2005: 150). Consequently, South Africa was paradoxically 'burdened' with a range of cultural heritage attractions that were high in quality yet distinctly unrepresentative of the nation as a whole.

As Bandyopadhyay *et al.* (2008: 800) have argued, nations containing histories that have been strongly influenced by periods of colonialism often seek to project an 'uncontaminated' image of the nation. However, the notion of 'contamination' – particularly in a South African sense – is a delicate issue. Indeed, it is often assumed that the contamination of a nation is *White* in nature, and as nations like India and Malaysia gained independence from British rule, they could reassert a non-White identity due to the relative absence of European settler populations. The situation in South Africa of course does not permit this. Although South Africa ceased to be a British colony in 1961, the nation retained a deeply embedded population of White South Africans. As Goudie *et al.* (1999: 22) observed, tourism development in South Africa was quickly identified as a 'catalyst for social change and healing in South Africa' early in the democratic era. However, the South African government has often struggled to develop an inclusive cultural heritage strategy and representative narratives for its diverse ethno-racial population. Instead, in an attempt to validly reassert an Afrocentric discourse in the nation's cultural heritage, the

ANC (African National Congress) governments that have successfully retained power since 1994 have frequently developed triumphalist histories that inadvertently alienate other sections of society (Deacon, 2000; Strange & Kempa, 2003).

In many contexts, the state will be responsible for which facets of cultural heritage receive greater attention (Burns, 2005; Cano & Mysyk, 2004; Johnson, 1999; Light, 2007). This is because cultural heritage may 'shape and control' current and future national identities (Palmer, 1999: 319). To some extent, political discourse has already influenced cultural heritage development in South Africa (Witz *et al.*, 2001). Indeed, Shackley's (2001) observations of Robben Island Museum (RIM) noted that whilst most visitors were keen to learn South African history, the museum failed to construct an objective view of Robben Island's political past. The influence of governmental bodies and political narratives may also foster 'social amnesia' (Park, 2011: 523). As Strange and Kempa (2003: 388–389) observe, RIM often projected experiences that removed 'oppositional stories' to an extent that 'onsite heritage interpreters and external pressure groups lobbied to introduce alternative and in some cases discordant strains into site interpretation ... so that more nuanced (and sometimes less optimistic) appraisals of the struggle for democracy are beginning to challenge triumphalist accounts'. Indeed, to counter overtly Eurocentric cultural heritage development with overtly Afrocentric doctrine is clearly not the correct solution for a democratic South Africa.

Redefining the role of Afrikaner cultural heritage sites in the post-apartheid era remains a prominent issue in particular. How such locations are developed along the lines of 'a more nuanced identity premised on cultural preservation and multiculturalism' has been difficult to establish (Autry, 2012: 154). Autry (2012) observed the juxtaposing of Pretoria's Afrocentric Freedom Park to the Eurocentric Voortrekker Monument as a notable example due to the contradictory narratives that were developed. Here it is important to note that whilst a rebalancing of cultural heritage is required, histories should not be 'rewritten' or edited to establish a new national consciousness and identity. The removal of problematic characters, events and even place names (see Goulding & Doric, 2009; Yeoh & Kong, 1996) should not be seen as an acceptable solution also. Indeed, it has been argued that Afrikaners continue to struggle to develop identities 'that could coherently integrate the Anglo-Boer war past, the apartheid past and the new South Africa present' (van der Waal & Robins, 2011: 772). Therefore, as Autry (2012: 147) suggests, although new 'national markers' are required, South Africa must also continue to engage with its inherited cultural heritage.

The limited visibility of cultural heritage products

The current vision of the NTSS (NDT, 2011c: 6) includes the repositioning of South Africa to be among the top 20 tourism destinations in the world by 2020. Unfortunately, the NTSS has frequently failed to identify the potential of cultural heritage tourism to achieve this goal, although it does acknowledge that cultural heritage attractions are of 'poor quality' and inadequately managed (NDT, 2011c: 22). The medium-term Strategic Plan (2011/2012–2015/2016) of the National Department of Tourism identified 'the development of niche products such as cultural heritage tourism as a priority' (NDT, 2011d: 6), yet cultural heritage tourism still remains largely invisible in South African government documentation relating to the tourism industry (NDT, 2012a: 5). During the apartheid era, cultural heritage had largely been constructed using a Eurocentric narrative that valued the settler cultures of the British and Dutch as superior to those that were African. Instead, African cultural heritage was often 'hidden from view' for many years (Phaswana-Mafuya & Haydam, 2005: 151). However, whilst attempting to foster 'a more diverse and sensitive portrayal of South African history' (Goudie et al., 1999: 24), including a shift in focus to rural cultural heritage and traditional cultures (van Veuren, 2004), it remains difficult to understand why cultural heritage tourism only appears sporadically in tourism growth strategies. Indeed, the visibility of South Africa's cultural heritage remains limited even in the post-apartheid era.

Despite its considerable potential to stimulate economic growth, the cultural heritage potential of South Africa frequently emerges on the periphery of debate in comparison to other forms of tourism. It also rarely achieves a consistent appraisal in strategies, particularly if one compares NDT and NTSS discourses. There is no doubt that political cultural heritage sites depicting the history of apartheid are highly popular tourist attractions – particularly those that are linked to the life of Nelson Mandela such as RIM, the Constitution Hill prison site in Johannesburg, Mandela's House Museum in Soweto and Mandela's homestead in Qunu, Eastern Cape. However, whilst the so-called *Mandela factor* has played an important role in tourism growth, successive South African governments have failed to develop viable strategies that maximise this potential.

The visibility of South Africa's cultural heritage attractions from a geographical perspective is also of particular concern. SAT marketing efforts are often in direct contrast with the aims of improving the geographical spread of tourism (SAT, 2010b: 18). Although SAT's Tourism Growth Strategy (2010b) has identified the need to foster a wider geographical coverage of tourism development as one of its six key objectives – more even distribution of international visitors throughout South Africa – 'its marketing efforts are in fact attracting visitors to very specific parts of the country' (Visser & Hoogendoorn, 2012: 72). Indeed, the promotion of cultural heritage attractions in the provinces of the Western Cape and Gauteng noticeably eclipse the methods and frequency of approaches designed to foster growth in other regions. This issue, of course, continues to further undermine government efforts to create jobs and alleviate poverty.

Poor experiential encounters

The experiential encounters of international tourists engaging with South African cultural heritage have frequently been reported to be of significant concern in recent SAT Growth Strategy documents (2008–2010 & 2011–2013) as well as in every Annual Tourism Report since 2007 (SAT, 2007a, 2008, 2009, 2010a, 2011, 2012). Indeed, it has twice been reported that 'foreign tourists are exposed to fewer and less authentic cultural experiences than they expect or desire' (SAT, 2007a: 27; SAT, 2010b: 36) and that '[South Africa's] cultural assets are largely unclear in the consumer's mind and undifferentiated from the rest of the continent' (SAT, 2007b: 37). Evidence to suggest that marketing and promotion of the country – and its range of new cultural heritage products – has also fared badly is supported by the finding that 'in terms of offerings, products, and experiences – in the mind of consumers globally, South Africa remains, on the whole, much the same as what it was 10 to 15 years ago' (SAT, 2010b: 46).

Although the vision of the National Heritage and Culture Tourism Strategy (NHCTS) is 'to realise the global competitiveness of South African heritage and cultural resources' (NDT, 2012a: 15), visitor experiences suggest that this potential remains largely unlocked. This issue, arguably, is due to a combination of the different factors mentioned in this chapter and the noticeable absence of adequate governmental approaches to record and analyse international tourists' (and domestic tourists, for that matter) experiences of cultural heritage. Indeed, SAT's exit questionnaire for tourists does not even provide an option for cultural heritage as a motive for travel. To date, reports have been sporadic in nature and have failed to provide sufficient levels of information to further explore why tourist experiences of cultural heritage fare badly. Moreover, the ability to monitor the volume and value of cultural heritage tourism in South Africa is noticeably limited. Without having access to information it is impossible for national government to drive cultural heritage strategies or make informed decisions about future developments and investments in cultural heritage tourism. Next, we conclude the chapter by presenting a range of potential solutions to some of the critical issues identified in this chapter.

Key Recommendations and Conclusions

Depoliticising South African cultural heritage

Cultural heritage attractions may act as mediators, and even points of reconciliation, in the formation of a new identity and it is crucial that South Africa finds a viable way to use its cultural heritage in a similar way. However, as previously discussed, in an attempt to realign the nation's identity to be more reflective of Afrocentric discourse, successive South African governments have failed to develop a solution that is reflective of its highly diverse cultural heritage. Although this is undoubtedly a considerable challenge, a particular criticism of previous approaches has been the use of cultural heritage attractions to act as vehicles for political narratives to be transmitted. While cultural heritage settings permit national identities to be developed (Akama, 2004; Desforges, 2000; Hall, 2000; Kim & Jamal, 2007; McCabe & Stokoe, 2004; Park, 2010, 2011; Yeoh & Kong, 1996) they may also act as stubborn barriers to cohesion also. Indeed, the development of cultural heritage that is selective or particular to certain sections of a society can lead to negative outcomes, including further fragmentation and isolation (Butler *et al.*, 2014; Goulding & Doric, 2009; McCrone, 2001; Palmer, 2005; Poria & Ashworth, 2009; Tunbridge & Ashworth, 1996; Yang *et al.*, 2006). It is crucial then, that if the new South Africa is to foster collective identities, overtly political discourses that perpetuate feelings of 'us' and 'them' must be distanced from cultural heritage attractions. In a similar vein, it is also crucial that South Africa's White cultural heritage remains an important feature of contemporary and future strategies to develop tourism further. Although painful memories of the apartheid era still resonate, it would be a considerable oversight to neglect such a vital section of the nation's cultural fabric. Although this is perhaps easier said than done, it is of significant importance that South African cultural heritage is representative of a nation rich in diversity.

The development of cultural heritage routes

In response to the narrow geographical focus of cultural heritage promotion, future strategies should be constructed to focus specifically on the development of cultural heritage routes. These routes may involve a range of different attractions including cultural villages, township tours and liberation heritage. A number of recent studies have revealed that heritage routes can act as key drivers for local economic developments (LEDs) and poverty alleviation in rural areas (Rogerson, 2007; Snowball & Courtney, 2010). Heritage routes also increase the geographical spread of tourism, and help promote rural cultural heritage that benefits local communities. This approach permits the combination of secondary attractions that exhibit limited pulling power, with primary attractions that act as key waypoints (Ivanovic, 2008: 157). A promising example is the proposed Liberation Heritage Route (LHR) project which has been designed to emerge as a potential World Heritage Site. The LHR project will be developed around heritage cluster sites that amalgamate UNESCO World Heritage Sites with over 200 other minor heritage attractions (Bialostocka, 2013: 1). However, despite these proposals it is interesting to note that route development is not mentioned in any current SAT strategies.

The establishment of a geographically diverse cultural heritage

The diversification of cultural heritage tourism products should be a two-fold strategy. First, it should be developed as a tool for innovation which will increase competitiveness and attraction quality (Booysen, 2012: 122). Second, it should act as a vehicle for employment generation and poverty alleviation and as it spreads the economic impacts of tourism to a wider range of destinations (Dickson, 2012; Rogerson, 2011: 204). Just two provinces, Gauteng and the Western Cape, accounted for 75% of all international overnight stays in 2012 (SAT, 2013: 107). This figure reveals the ineffectiveness of the current strategy in activating rural cultural heritage tourism in other provinces, such as Mpumalanga and the Eastern Cape. South Africa's *New Growth Path* has identified tourism (EDD, 2011: 26) as a priority sector expected to create 225,000 direct jobs (EDD, 2011: 65; Manyathi, 2012: 26) by 2020 using the National Culture and Heritage Tourism Strategy (NDT, 2011b, 2012a) and the Rural Tourism Strategy (NDT, 2011a) as the main drivers. Although this is a logical approach, the need to enhance cultural and heritage tourism products in the marketing of South Africa as a destination was raised as an important issue (NDT, 2012a: 10). Similarly, the National Craft Sector Development Programme (DAC, 2011: 8) was constructed to create jobs in rural areas via a combination of the creative industries with tourism. However, the programme appears unrealistic if South Africa fails to diversify the range of cultural heritage attractions international tourists engage with (van Beek & Schmidt, 2012).

Researching and benchmarking cultural heritage performance

To date, the South African Heritage and Culture Tourism Strategy has failed to improve the performance of cultural heritage tourism attractions in

South Africa. A key reason for this failure is perhaps attributed to limited research into understanding why many international tourists remain unsatisfied with cultural heritage experiences. New research programmes are urgently required to explore the characteristics and experiential demands of an increasingly volatile international tourist market. After all, South Africa is blessed with an exceptional portfolio of diverse cultural heritage attractions – however, it remains to be seen whether the country can maximise its potential in the future.

References

Akama, J.S. (2004) Neocolonialism, dependency and external control of Africa's tourism industry: A case study of wildlife safari tourism in Kenya. In C.M. Hall and H. Tucker (eds) *Tourism and Postcolonialism: Contested Discourses, Identities and Representations* (pp. 140–52). Oxford: Routledge.

Alvarez, M.D. and Korzay, M. (2011) Turkey as a heritage tourism destination: The role of knowledge. *Journal of Hospitality Marketing & Management* 20 (3–4), 425–440.

ATLAS (Association for Tourism and Leisure Education) (2007) *ATLAS Cultural Tourism Survey: Summary Report*. The Netherlands: Arnhem.

ATLAS (Association for Tourism and Leisure Education) (2009) Experiencing difference: Changing tourism and tourist experiences. ATLAS Reflections, 2009 series. The Netherlands: Arnhem.

Autry, R.C. (2012) The monumental reconstruction of memory in South Africa: The Voortrekker Monument. *Theory, Culture & Society* 29 (6), 146–164.

Bandyopadhyay, R., Morais, D.B. and Chick, G. (2008) Religion and identity in India's heritage tourism. *Annals of Tourism Research* 35 (3), 790–808.

Bialostocka, O. (2013) Liberation Heritage Route: Reminiscent of the painful past or a road to the future? (AISA Policy Brief Number 100). Africa Institute of South Africa.

Binns, T. and Nel, E. (2002) Tourism as a local development strategy in South Africa. *The Geographical Journal* 168 (3), 235–247.

Booysen, I. (2012) Innovation in tourism: A new focus for research and policy development in South Africa. *Africa Insight* 42 (2), 112–126.

Bredekamp, H.C. (2006) Transforming representations of intangible heritage at Iziko (national) museums, South Africa. *International Journal of Intangible Heritage* 1 (1), 76–82.

Briedenhann, J. and Wickens, E. (2004) Tourism routes as a tool for the economic development of rural areas – vibrant hope or impossible dream. *Tourism Management* 25, 71–79.

Burns, P. (2005) Social identities, globalisation and the cultural politics of tourism. In W. Theobald (ed.) *Global Tourism* (3rd edn) (pp. 391–405). Amsterdam: Elsevier.

Butler, G., Khoo-Lattimore, C. and Mura, P. (2014) Heritage tourism in Malaysia: Fostering a collective national identity in an ethnically diverse country. *Asia Pacific Journal of Tourism Research* 19 (2), 199–218.

Cano, L.M. and Mysyk, A. (2004) Cultural tourism, the state, and day of dead. *Annals of Tourism Research* 31 (4), 879–898

Coombes, A.E. (2003) *History after Apartheid: Visual Culture and Public Memory in a Democratic South Africa*. Durham, NC: Duke University Press.

DAC (Department of Arts and Culture of the Republic of South Africa) (2011) *National Craft Sector Development Programme*. Pretoria: DAC.

Deacon, H. (2000) Memory and history at Robben Island. Paper presented at memory and history conference, University of Cape Town (unpublished).
DEAT (Department of Environmental Affairs and Tourism) (1996) *White Paper: The Development and Promotion of Tourism in South Africa*. Pretoria: Government of South Africa.
DEAT and DTI (Department of Environmental Affairs and Tourism and Department of Trade and Industry) (2004) *Global Competitiveness Project, Phase 1*. Pretoria: DEAT and DTI.
Desforges, L. (2000) Travelling the world – identity and travel biography. *Annals of Tourism Research* 27 (4), 926–945.
Dickson, J.L. (2012) Revisiting 'township tourism': Multiple mobilities and the re-territorialisation of township spaces in Cape Town, South Africa. *Anthropology Southern Africa* 35 (1–2), 31–41.
EDD (Economic Development Department) (2011) *New Growth Path: Framework*. Pretoria: Economic Development Department.
Garrod, B. and Fyall, A. (2000) Heritage tourism: A question of definition. *Annals of Tourism Research* 28 (4), 1049–1052.
Garrod, B. and Fyall, A. (2001) Managing heritage tourism. *Annals of Tourism Research* 27 (3), 682–708.
Gössling, S. (2000) Sustainable tourism development in developing countries: Some aspects of energy use. *Journal of Sustainable Tourism* 8 (5), 410–425.
Goudie, S.C., Khan, F. and Kilian, D. (1999) Transforming tourism: Black empowerment, heritage and identity beyond apartheid. *South African Geographical Journal* 81 (1), 22–31.
Goulding, C. and Doric, D. (2009) Heritage, identity and ideological manipulation: The case of Croatia. *Annals of Tourism Research* 36 (1), 85–102.
Hall, C.M. (2000) *Tourism Planning*. Harlow: Prentice Hall.
Hughes, H. and Allen, D. (2005) Cultural tourism in Central and Eastern Europe: The views of 'induced image formation agents'. *Tourism Management* 26 (2), 173–183.
Ivanovic, M. (2008) *Cultural Tourism*. Cape Town: Juta Academic.
Ivanovic, M. and Saayman. M. (2013) South Africa calling cultural tourists. *African Journal for Physical, Health Education, Recreation and Dance* Supplement 2 (September), 138–154.
Johnson, N.C. (1999) Framing the past: Time, space and the politics of heritage tourism in Ireland. *Political Geography* 18, 187–207.
Kim, H. and Jamal, T. (2007) Touristic quest for existential authenticity. *Annals of Tourism Research* 34 (1), 181–201.
Lepp, A. and Gibson, H. (2011) Reimaging a nation: South Africa and the 2010 FIFA World Cup. *Journal of Sport & Tourism* 16 (3), 211–230.
Light, D. (2001) 'Facing the future': Tourism and identity-building in post-socialist Romania. *Political Geography* 20 (8), 1053–1074.
Light, D. (2007) Dracula tourism in Romania: Cultural identity and the state. *Annals of Tourism Research* 34 (3), 746–765.
Manyathi, O. (2012) Growing the tourism sector. *Public Sector Manager* September, 26–29.
Massyn, P.J. and Koch, E. (2004) African game lodges and rural benefit in two southern African countries. In C.M. Rogerson and G. Visser (eds) *Tourism and Development Issues in Contemporary South Africa* (pp. 102–138). Pretoria: Africa Institute of South Africa.
McCabe, S. and Stokoe, E.H. (2004) Place and identity in tourist accounts. *Annals of Tourism Research* 31 (3), 601–622.

McCrone, D. (2001) *Understanding Scotland: The Sociology of the Nation*. London: Routledge.
McIntosh, A. and Prentice, R. (1999) Affirming authenticity: Consuming cultural heritage. *Annals of Tourism Research* 26 (3), 589–612.
McKercher, B. (2002) Towards a classification of cultural tourists. *International Journal of Tourism Research* 4, 29–38.
Ndlovu, N. and Rogerson, C.M. (2004) The local economic impacts of rural community-based tourism in the Eastern Cape. In C.M. Rogerson and G. Visser (eds) *Tourism and Development Issues in Contemporary South Africa* (pp. 436–451). Pretoria: Africa Institute of South Africa.
NDT (National Department of Tourism) (2011a) *National Rural Tourism Strategy*. Pretoria: National Department of Tourism.
NDT (National Department of Tourism) (2011b) *National Strategy on Heritage Tourism (Draft)*. Pretoria: National Department of Tourism.
NDT (National Department of Tourism) (2011c) *National Tourism Sector Strategy (NTSS)*. Pretoria: National Department of Tourism.
NDT (National Department of Tourism) (2011d) *Medium Term Strategic Plan 2011/2012–2015/2016*. Pretoria: National Department of Tourism.
NDT (National Department of Tourism) (2012a) *National Strategy on Heritage and Culture Tourism (Final)*. Pretoria: National Department of Tourism.
NDT (National Department of Tourism) (2012b) *The Experiences, Perceptions and Expectations of Consumers of Cultural Tourism Products: An Exploratory Study*. Pretoria: National Department of Tourism.
OECD (Organisation for Economic Cooperation and Development) (2009) *The Impact of Culture on Tourism*. Paris: OECD.
Palmer, C. (1999) Tourism and the symbols of identity. *Tourism Management* 20 (3), 313–321.
Palmer, C. (2005) An ethnography of Englishness: Experiencing identity through tourism. *Annals of Tourism Research* 32 (1), 7–27.
Park, H. (2010) Heritage tourism: Emotional journeys into nationhood. *Annals of Tourism Research* 37 (1), 116–135.
Park, H. (2011) Shared national memory as intangible heritage: Re-imagining two Koreas as one nation. *Annals of Tourism Research* 38 (2), 520–539.
Phaswana-Mafuya, N. and Haydam, N. (2005) Tourists' expectations and perceptions of the Robben Island Museum – A world heritage site. *Museum Management and Curatorship* 20 (2), 149–169.
Pine, B.J. and Gilmore, J.H. (1999) *The Experience Economy: Work is Theatre & Every Business a Stage*. Boston, MA: Harvard Business School Press.
Poria, Y. and Ashworth, G. (2009) Heritage tourism – current resource for conflict. *Annals of Tourism Research* 36 (3), 522–525.
Pretes, M. (2003) Tourism and nationalism. *Annals of Tourism Research* 30 (1), 125–142.
Richards, G. (2007) *Cultural Tourism: Global and Local Perspective*. New York: Haworth Hospitality Press.
Rogerson, C.M. (2007) Tourism routes as vehicles for economic development in South Africa: The example of the Magaliesberg Meander. *Urban Forum* 18, 49–68.
Rogerson, C.M. (2009) Tourism development in Southern Africa: Patterns, issues and constrains. In J. Saarinen, F. Becker, H. Manwa and D. Wilson (eds) *Sustainable Tourism in Southern Africa: Local Communities and Natural Resources in Transition* (pp. 20–41). Bristol: Channel View Publications.
Rogerson, C.M. (2011) Niche tourism policy and planning: The South African experience. *Tourism Review International* 15 (1–2), 199–211.

Saarinen, J., Becker, F., Manwa, H. and Wilson, D (2009) Introduction: Call for sustainability. In J. Saarinen, F. Becker, H. Manwa and D. Wilson (eds) *Sustainable Tourism in Southern Africa: Local Communities and Natural Resources in Transition* (pp. 3–19). Bristol: Channel View Publications.

SAT (South African Tourism) (2007a) *2006 Annual Tourism Report*. Pretoria: South African Tourism Strategic Research Unit.

SAT (South African Tourism) (2007b) *Tourism Growth Strategy 2008–2010* (3rd edn). Pretoria: South African Tourism Strategic Research Unit.

SAT (South African Tourism) (2008) *2007 Annual Tourism Report*. Pretoria: South African Tourism Strategic Research Unit.

SAT (South African Tourism) (2009) *2008 Annual Tourism Report*. Pretoria: South African Tourism Strategic Research Unit.

SAT (South African Tourism) (2010a) *2009 Annual Tourism Report*. Pretoria: South African Tourism Strategic Research Unit.

SAT (South African Tourism) (2010b) *The Marketing Tourism Growth Strategy for South Africa 2011–2013*. Pretoria: South African Tourism Strategic Research Unit.

SAT (South African Tourism) (2011) *2010 Annual Tourism Report*. Pretoria: South African Tourism Strategic Research Unit. South Africa.

SAT (South African Tourism) (2012) *2011 Annual Tourism Report*. Pretoria: South African Tourism Strategic Research Unit.

SAT (South African Tourism) (2013) *2012 Annual Tourism Report*. Pretoria: South African Tourism Strategic Research Unit.

Shackley, M. (2001) Potential futures for Robben Island: Shrine, museum or theme park? *International Journal of Heritage Studies* 7 (4), 355–363.

Snowball, J.D. and Courtney, S. (2010) Cultural heritage routes in South Africa: Effective tools for heritage conservation and local economic development? *Development Southern Africa* 27 (4), 563–576.

Strange, C. and Kempa, M. (2003) Shades of dark tourism: Alcatraz and Robben Island. *Annals of Tourism Research* 30 (2), 386–405.

Tunbridge, J.E. and Ashworth, G.J. (1996) *Dissonant Heritage: The Management of the Past as a Resource in Conflict*. New York: Wiley.

van Beek, W. and Schmidt, A. (2012) African dynamics of cultural tourism. In W. van Beek and A. Schmidt (eds) *African Hosts and their Guests: Cultural Dynamics of Tourism* (pp. 1–33). Suffolk: Boydell & Brewer.

van der Ark, L.A. and Richards, G. (2006) Attractiveness of cultural activities in European cities: A latent class approach. *Tourism Management* 27 (6), 408–1413.

van der Waal, K and Robins, S. (2011) 'De la Rey' and the revival of 'Boer Heritage': Nostalgia in the post-apartheid Afrikaner culture industry. *Journal of Southern African Studies* 37 (4), 763–779.

van Veuren, E.J. (2004) Cultural village tourism in South Africa: Capitalizing on indigenous culture. In C.M. Rogerson and G. Visser (eds) *Tourism and Development Issues in Contemporary South Africa* (pp. 39–160). Pretoria: Africa Institute of South Africa.

Visser, G. and Hoogendoorn, G. (2012) Uneven tourism development in South Africa. *Africa Insight* 42 (2), 66–74.

Witz, L., Rassol, C. and Minkley, G. (2001) Repackaging the past for South African tourism. *Daedalus* 130 (1), 277–296.

Woodward, K. and Goldblatt, D. (2011) Introduction. *Soccer & Society* 12 (1), 1–8.

Yeoh, B. and Kong, L. (1996) The notion of place in the construction of history, nostalgia and heritage in Singapore. *Singapore Journal of Tropical Geography* 17 (1), 52–65.

6 Cultural Tourism and the Arts Festivals

Corné Pretorius

Our everyday life is someone else's adventure
Wales Tourist Board (2005: 5)

Introduction

Culture can be seen as a main pull factor for travellers when they make a decision to travel. It is then not surprising to find that cultural attractions have become important in the development of tourism (Akama, 2002; Reid, 2002; Richards, 2001). This is also the case for tourism development in South Africa, where cultural attractions have long played a role in the development of tourism in the country (SATourism, 2014: online). Travellers are captivated by the diverse cultures in the country, as this mix of cultures, with African, European and Asian influences, is fused to create a unique South African multicultural society (Ramchander, 2007: 39). Thus, the country has a comparative advantage for the development of cultural tourism when reference is made to the unique attractions of these cultures as these are the attractions sought by travellers (Akama, 2002).

Cultural Tourism

Cultural tourism has continued to be a major growth industry in the world (Richards, 2002; Smith, 2009). Definitions of cultural tourism are broadening and changing all the time as this sector of tourism is becoming more diverse (Richards, 2007; Smith, 2009). Primarily, cultural tourism can be thought of as the component of tourism oriented towards the culture of a particular country or community (Anheier & Isar, 2008; Ivanovic, 2008). Cultural tourism can also be thought of as tourism constructed, proffered and consumed explicitly or implicitly as cultural appreciation, either as experiences or through schematic knowledge gaining (Prentice, 2001). Richards (2007) further defines cultural tourism in two ways. First, that cultural tourism is all movements of persons to specific cultural attractions such as museums, heritage sites, artistic performances and festivals outside their normal place of residence (the technical definition) and, second, cultural tourism is the movement of persons to cultural manifestations away from their normal place of residence, with the intention to gather new information and experiences to satisfy their cultural needs (the conceptual definition). Cultural tourism is then the passive, active and interactive engagement with culture(s) and communities, whereby the visitor gains new experience of an educational, creative and/or entertaining nature (Richards, 2007; Smith, 2009).

Cultural tourism can then be explained within the context of culture (Akama, 2002: 14). Culture serves the purpose of teaching one how to do things (i.e. way of life, artworks and cultural products) and how to think (i.e. attitudes, ideas, values and beliefs) to communicate and to organise the world (Anheier & Isar, 2008; Akama, 2002; Ivanovic, 2008; Reid, 2002; Reisinger, 2009; Reisinger & Turner, 2012). As such, culture then includes several considerations, as suggested by Reisinger and Turner (2012) and Richards (2001). Some of these considerations include that: culture is the lived and creative experience of individuals; culture is a body of artefacts, symbols, texts and objects; culture involves enactment and representation; culture embraces the arts and art discourses; culture is the commoditised output of the cultural industries; culture is the expressions of everyday life; and culture is learned from other members of the society.

Culture is then the sum of acquired experience and knowledge, and so produces a wide range of cultural products for tourism consumption (Briedenhann & Wickens, 2007). Culture itself can also be seen as a product when viewed as the intellectual and artistic works and practices of individuals or groups to which certain specific meanings are attached (Ivanovic, 2008). Cultural products are produced for various reasons. According to Kumphai (2000) and Ivanovic (2008), cultural products include textiles, wood, ceramics, glass and metal; they embody aesthetic features and production technologies that are deeply enmeshed in each producer's local traditions. Lee (2002) and Kumphai (2000) state that cultural products include what are typically called handcrafts, as some level of hand production is common. Cultural products can then be tangible or intangible creations of a particular culture. Tangible cultural products include paintings, pottery, artwork, a cathedral, a piece of literature, or a pair of chopsticks. Intangible cultural products include an oral tale, a dance, music, a sacred ritual, a system of education, or a law. In summary, cultural products are goods and services that include the arts (performing arts, visual arts, architecture), heritage conservation (museums, galleries, libraries), the cultural industries (written media, broadcasting, film, recording), and festivals. Whatever the form of cultural product, its presence within the culture is required or justified by the underlying beliefs, attitudes, ideas and values of that culture (Ivanovic, 2008; Kumphai, 2000). Thus, cultural products reflect a culture's perspectives or view of the world.

With this in mind, cultural tourism then covers not just the consumption of the cultural products of the past, but also the contemporary culture of a people or a region, and visiting cultural sites has tended to dominate the development of cultural tourism (Akama, 2002; Richards, 2001).

Cultural tourism can then be characterised on its typology (Reid, 2002). Smith (2009) provides a fairly comprehensive typology of the attractions of cultural tourism (Smith, 2009). These researchers state that cultural tourism includes heritage tourism (which includes visits to castles, palaces, country houses, archaeological sites, monuments, museums, architecture, religious sites, etc.), arts tourism (which includes visits to the theatre, concerts, galleries, festivals, carnivals, events, literary sites, etc.), creative tourism (e.g. photography, painting, pottery, dance, cookery, crafts and creative industries such as film, TV, fashion and design) and indigenous tourism (which includes hill tribe, desert, jungle, rainforest or mountain trekking, tribal villages, visits to cultural centres, arts and crafts, cultural performances, festivals, etc.). In summary, cultural tourism can be seen as covering both heritage tourism (related to artefacts of the past) and arts tourism (related to contemporary cultural production) (Akama, 2002: 15; Richards, 2001). From this, cultural tourism is indeed broad in its remit (see Figure 6.1).

Arts Tourism

Arts tourism has become one of the flows through which cultural exchange can take place (Richards, 2007). Richards (2007) also states that the increasing scale of such exchanges in turn becomes a stimulus for cultural tourism. Thus, arts tourism has become a component of cultural tourism (Plangpramool, 2013). According to several national and international researchers, the arts – i.e. culture perspective or practice – have become a new motivation for travel (see Goeldner & Ritchie, 2009; Plangpramool, 2013; Quinn, 2006). Arts tourism can then be defined as the travel for the purpose of experiencing the elements of culture (Hughes, 2012; Ivanovic, 2008). It also gives way to the establishment of arts festivals, which host a variety of arts and culture products for the festival visitors to enjoy.

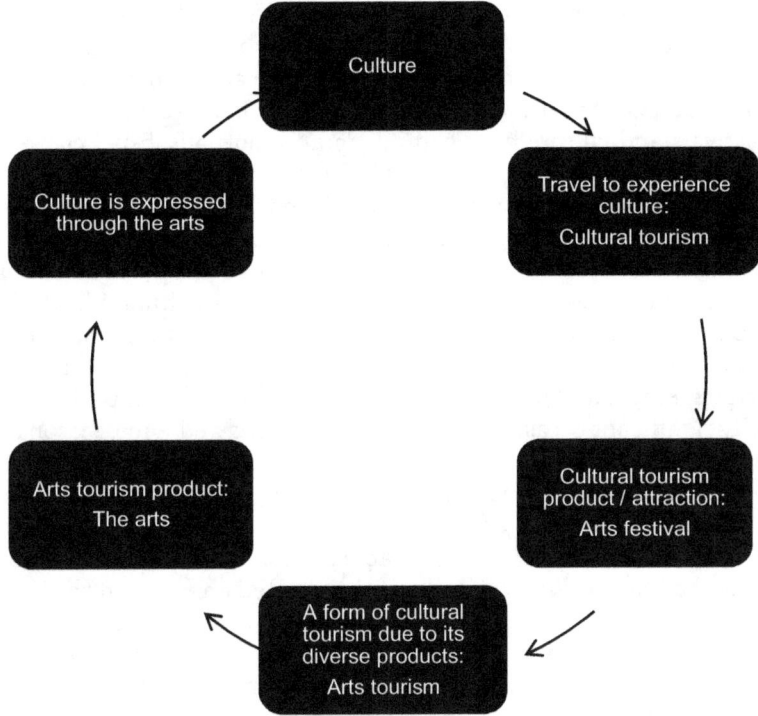

Figure 6.1 Relationship between cultural tourism and the arts festival
Source: Author's own adaptation.

The arts festival as cultural attraction

There exist a wide range of cultural attractions around the world; festivals can be seen as one of these attractions (Richards, 2001). Festivals can be viewed as products that are desired by tourists (Picard & Robinson, 2006), providing a whole range of colourful exhibitions or spectacles to observe, admire or participate in (Smith, 2009: 121). Festival visitors primarily participate in festivals because of a special interest in the product, event, heritage or culture being celebrated (Zeppel & Hall, 1992). Thus, a festival can be described as a community-themed event or celebration designed to display different art forms and activities, along with related tourism and hospitality experiences (Kruger & Petzer, 2008). Festivals have become a cultural phenomenon (Smith, 2009) as the generic origins of a festival have cultural meaning to the host community (Arcodia & Whitford, 2006) and many such events serve as an expression of the culture of a particular community

(Smith, 2009). This is true among the cultures of South Africa (Hauptfleisch, 2007). As a result, festivals have become a subject of interest in cultural studies where they are often incorporated in the literature on cultural tourism (see Getz, 2008; Ivanovic, 2008; Quinn, 2006; Richards, 2001; Smith, 2009).

Festivals and culture have had a long and mutually beneficial history (Smith, 2009). They can be viewed as cultural drivers (Hauptfleisch, 2007). The reasons why festivals were established, and why they can be viewed as cultural products are highlighted by Smith (2009) and Zeppel and Hall (1992). Festivals offer: a means of reaffirming or reviving a local culture or tradition; communities the chance to celebrate their cultural identity; support to local artists; a concentrated period of high-quality artistic activity; and an opportunity to celebrate dance, drama, comedy, film and music, the arts, crafts, ethnic and indigenous cultural heritage, religious traditions, historically significant occasions, sporting events, food and wine, seasonal rites and agricultural products.

Festivals can take numerous forms (Kruger & Saayman, 2012; Smith, 2009), for example: beverage festivals (including wine festivals, beer festivals, coffee festivals and liquor festivals – an example of such a festival in South Africa is the Wacky Wine Festival which is hosted annually in the Western Cape Province) (Lyck *et al.*, 2012; Yuan *et al.*, 2005); agricultural festivals (the celebration of new land produce, harvest and/or cattle and may include any form of food festivals – examples include the Cherry Festival and the Cheese Festival, all hosted in the Western Cape Province) (Farlex, 2014; Hall & Sharples, 2008); community festivals (including children's festivals and festivals for disabled people) (Lyck *et al.*, 2012); sport festivals (held with the purpose of competition – for example, the Nelson Mandela Bay Water Festival) (Lyck *et al.*, 2012; Mair, 2009); vehicle festivals (including air shows, motorbike festivals and bicycle festivals) (Lyck *et al.*, 2012); technology festivals (including computer festivals) (Lyck *et al.*, 2012); fashion festivals (Lyck *et al.*, 2012); historic festivals (Lyck *et al.*, 2012); religious festivals (having religious significance) (Arcodia & Robb, 2000; Farlex, 2014); and arts festivals (including sculpture festivals, painting festivals, dancing festivals, literature festivals, theatre festivals, antique festivals, glass festivals, opera festivals, musical festivals, film festivals, ethnicity festivals, and sewing, knitting, weaving and needlework festivals – for example the Klein-Karoo National Arts Festival, the Aardklop National Arts Festival, Gramstown Arts Festival and the Oppikoppi Music Festival) (Farlex, 2014; Inkei, 2005; Kruger & Saayman, 2012). The focus of this chapter is the arts festival.

Arts festivals have been widely researched (see Getz, 2008; Hauptfleisch, 2004; Hughes, 2012; Korza & Magie, 1998; Quinn, 2005, 2006; Tassiopoulos, 2010). An arts festival can be defined as a festival that presents, over a short period, a variety of arts and cultural works or products created or produced by other professional organisations or artists working in diverse artistic disciplines (Inkei, 2005). Faulkner *et al.* (2001), Hughes (2012) and Korza and Magie (1989) identified reasons for the establishment of the arts festival which further support the notion that an arts festival can be seen as a cultural product or attraction. These reasons may include: a desire to celebrate; a desire to promote awareness of and to increase understanding of a particular art or cultural form; enabling arts and culture attendance by local residents where there are few other opportunities; encouraging community coherence through arts participation; encouraging a desire to exchange artistic and cultural ideas; interacting with people who share common or different values and lifestyles; and renewing cultural ties within the community.

The arts festival then aims to celebrate culture through a diverse range of the arts. The arts are then a reflection of culture and forms of interaction with other cultures (Phillips & Steiner, 1999). The arts can also be used to verbally and non-verbally communicate the culture of a particular community (Reisinger & Turner, 2012). The arts are commonly used to describe something of beauty, or a skill which produces an aesthetic result and requires some kind of creative impulse (Craig, 2014). Farlex (2014) defines art as the conscious use of the imagination in the production of objects intended to be contemplated or appreciated as beautiful, as in the arrangement of forms, sounds or words. The arts include numerous characteristics (see Farlex, 2014; Iavnovic, 2008; Phillips & Steiner, 1999; Zijlmans & Van Damme, 2008); the arts are free creation; unfettered by functional requirements; the imitation of abstract ideas or natural objects; every work of art figures out a new law/a new way of looking at the world; the arts express meaningfully local experiences and processes; convey a message of regional and national ethnicity; are a means of self-expression and communication; are rare, elite, original and costly; are closely related to concepts of beauty, skill, creativity, imagination, or self-expression; are composed of autonomous objects (paintings, sculptures, ceramics), or activities (dances, songs, performances); are the province of specialist artists; have no limitations; and cannot be restricted by ability, age or cultural background.

Further, the arts can be divided into two types: the performing arts and the visual arts (Edginton *et al.*, 2004; Esaak, 2014; Hughes, 2012; Kotler & Scheff, 1997). The performing arts include forms such as live music productions, drama (or theatre) and dance productions (Heilbrun & Gray, 2001; Kotler & Scheff, 1997; Whithers, 1980). The performing arts are those types of arts or creative activity being performed in public (Esaak, 2014; Farlex, 2014). The visual arts include such forms as literature, paintings, sculpture and crafts (Chaffee, 1984; Craig, 2014; Edginton *et al.*, 2004: 214; Esaak, 2014; Kendzulak, 2014) and can be described as decorative arts (Edginton *et al.*, 2004: 219; Esaak, 2014). Farlex (2014) defines the visual arts as objects that typically exist in permanent form and are primarily created through visual perception. In essence, the visual arts give voice and language to the otherwise mute art object (Chaffee, 1984). Different arts can be present at arts festivals (Snowball, 2010). Each arts festival in South Africa hosts a unique blend of arts of a particular culture as there are no particular forms of art or culture that are specific to tourism (Plangpramool, 2013).

The arts festival may then be viewed as a contributor to culture and cultural tourism – specific to the community hosting such events (see studies by Hauptfleisch, 2007; Korza & Magie, 1989). These contributions may be the celebration of the arts and culture of a community; the provision of entertainment through the arts to the enjoyment of a culture; the raising of money for arts and culture; the creation of a forum for cultural experimentation; the fostering of pride and commitment in the community; and the development of audiences for the arts and culture.

However, there are also a number of problems associated with the arts festival in its contribution to culture and cultural tourism (see Hauptfleisch, 2007; Hughes, 2012; Reisinger, 2009). Many communities may experience cultural changes due to tourism development (Reisinger, 2009). Culture is vulnerable, in the sense that art and culture that is dependent of an arts festival will prosper as long as the arts festival occurs (Hughes, 2012). Trivialisation of culture can occur in the process of attracting tourists. This may create pressure to produce 'popular' products, which may lead to the arts and culture product becoming commercialised and losing authenticity (Hughes, 2012). In addition, this may also cause erosion of local culture, as the growing demand for arts and culture products can stimulate mass-produced forms and can ultimately lower the quality of traditional artistic designs and forms, and symbolic values and meaning can disappear (Reisinger, 2009). Cultural commodification and transformation may then take place, as traditional culture is being packaged and treated as a commodity for sale to tourists (Reisinger, 2009). Culture diffusion can occur due to the arts festival. Cultural diffusion refers to the spread of cultural elements such as ideas and styles between individuals and groups within a single culture or from one culture to another. Cultural diffusion changes local cultural features and traditions (Reisinger, 2009), leading to cultural change within the community (Reisinger, 2009). Arts festivals cannot always fulfil the needs of all stakeholders attending an event and may become irrelevant for tourists as they may not understand local cultural significance (Hughes, 2012).

Conclusion

The development of cultural tourism in South Africa should take into consideration the role of cultural events, such as arts festivals. Cultural events are arenas in which forms of culture and the arts are displayed for visitors to enjoy, and, in return, they create exposure for a community and can stimulate the movement of travellers to these attractions.

In South Africa, each community holds its own form of cultural event to celebrate a unique aspect of that community. Tourists experience the culture being portrayed through the cultural and arts products on show at the different events (Wickens, 2005). Thus, the development of cultural and arts-related events can contribute to the flourishing of cultural tourism in South Africa as they celebrate diverse cultures that travellers would like to see (Anheier & Isar, 2008; Ivanovic, 2008; Richards, 2001; Smith, 2009).

References

Akama, J. (2002) Introduction: Cultural tourism in Africa. In J. Akama and P. Sterry (eds) *Cultural Tourism in Africa: Strategies for the New Millennium* (pp. 13–18). Conference Proceedings: The ATLAS Africa International Conference. Mombasa: Kenya. Retrieved from http://www.atlas-euro.org/pages/pdf/Cultural%20tourism%20in%20Africa%20Deel%201.pdf.

Anheier, H. and Isar, Y.R. (2008) *The Cultural Economy: The Cultures and Globalization* Series 2. London: Sage Publications.

Arcodia, C. and Robb, A. (2000) A future for events management: A taxonomy of event management terms. In J. Allen, R. Harris, L.K. Jago and A.J Veal (eds) *Conference Proceedings: Events Beyond 2000 – Setting the Agenda*. Conference on Event Evaluation, Research and Education (pp. 154–160). Sydney: University of Technology, School of Leisure, Sport and Tourism, Australian Centre for Event Management, 13–14 July.

Arcodia, C. and Whitford, M. (2006) Festival attendance and the development of social capital. *Journal of Convention & Event Tourism* 8 (2), 1–18. DOI:10.1300/J452v08n02_01.

Briedenhann, J. and Wickens, E. (2007) Developing cultural tourism in South Africa: Potential and pitfalls. In G. Richards (ed.) *Cultural Tourism: Global and Local Perspectives* (pp. 69–90). New York: Haworth Hospitality Press.

Chaffee, D. (1984) Visual art in literature: The role of time and space in ekphrastic creation. *Revista Canadiense De Estudicos* 8 (3), 311–320.

Craig, M. (2014) Meaning and definition of art. See www.visual-arts-cork.com/art-definition.htm#whatisart (accessed 14 February 2015).

Edginton, C.R., Hudson, S.D., Dieser, R.B. and Edginton, S.R. (2004) *Leisure Programming: A Service-centred and Benefit Approach* (4th edn). New York: McGraw-Hill.

Esaak, S. (2014) What are the visual arts? See www.arthistory.about.com/cs/reference/f/visual_arts.htm (accessed 15 January 2015).

Faulkner, B., Moscardo, G. and Laws, E. (2001) *Tourism in the 21st Century: Lessons from Experience*. London: Cromwell Press.

Free dictionary (Farlex) (2014) Food festivals, religious festivals, arts festivals, performing arts. See www.thefreedictionary.com/stratified+sampling (accessed 15 March 2015).

Getz, D. (2008) Event tourism: Definition, evolution, and research. *Tourism Management* 29 (3), 403–428. DOI:10.1016/j.tourman.2007.07.017.

Goeldner, C.R. and Ritchie, J.R.B. (2009) *Tourism: Principles, Practices, Philosophies* (11th edn). Hoboken, NJ: Wiley.

Hall, C.M. and Sharples, L. (2008) *Food and Wine Festivals: Development, Management and Markets*. Oxford: Butterworth-Heinemann.

Hauptfleisch, T. (2004) Eventification: Using the theatrical system to frame the event. In V. Cremona, P. Eversmann, H. Van Maanen, W. Sauter and J. Tulloch (eds) *Theatrical Events: Boarders, Dynamics, Frames* (pp. 279–302). Amsterdam: International Featre Research (IFTR).

Hauptfleisch, T. (2007) *Festivalising! Theatrical Events, Politics and Culture*. Netherlands: Rodopi.

Heilbrun, J. and Gray, C.M. (2001) *The Economics of Art and Culture* (2nd edn). Cambridge: Cambridge University Press.

Hughes, H.L. (2012) *Arts, Entertainment and Tourism*. Oxford: Butterworth-Heinemann.

Inkei, P. (2005) Assistance to arts and culture festivals: D'art topics in arts policy, 21. IFACCA.

Ivanovic, M. (2008) *Cultural Tourism*. Cape Town: Juta.

Kendzulak, S. (2014) Fine arts. See www.fineart.about.com/od/Glossary_F/g/Fine-Arts.htm (accessed 15 March 2015).

Korza, P. and Magie, D. (1989) *Arts Festival Work Kit*. Amherst: University of Massachusetts, Arts Extension Service.

Kotler, P. and Scheff, J. (1997) *Standing Room Only: Strategies for Marketing the Performing Arts*. Boston, MA: Harvard Business Press.

Kruger, S. and Petzer, D.J. (2008) Measuring tourists' satisfaction with quality of life issues at an arts festival. *ActaCommercii* 8 (1), 113–127.

Kruger, M. and Saayman, M. (2012) When do festinos decide to attend an arts festival? An analysis of the Innibos National Arts Festival. *Journal of Travel & Tourism Marketing* 29 (2), 147–162. DOI:10.1080/10548408.2012.648538.

Kumphai, P. (2000) Cultural products: Definition and website evaluation. Kasetsart University, Thailand, p. 145. See www.digital.library.okstate.edu/etd/umi-okstate-1979.pdf (accessed 25 January 2015).

Lee, S.E. (2002) Shopping for cultural products on the internet. MA dissertation, Iowa State University, Ames.

Lyck, L., Long, P. and Grige, A.X. (2012) *Tourism, Festivals and Cultural Events in Times of Crisis*. Frederiksberg Bogtrykkeri, Denmark: Copenhagen Business School Publications.

Mair, J. (2009) The events industry: The employment context. In T. Baum, M. Deer, C. Hanlon, L. Lockstone and K. Smith (eds) *People and Work in Events and Conventions: A Research Perspective* (pp. 3–16). Wallingford: CABI Publishing.

Phillips, R.B. and Steiner, C.B. (1999) *Unpacking Culture*. London: University of California Press.

Picard, D. and Robinson, M. (2006) *Festivals, Tourism and Social Change: Remaking Worlds*. Clevedon: Channel View Publications.

Plangpramool, S. (2013) Perceptions of selected attributes in tourism management of music festivals: A case study of Pattaya Music Festival 2012. *Asia-Pacific Journal of Innovation in Hospitality and Tourism* 2 (1), 53–67.

Prentice, R. (2001) Experiential cultural tourism: Museums and the marketing of the new romanticism of evoked authenticity. *Museum Management and Curatorship* 19 (1), 5–26. DOI:10.1080/09647770100201901.

Quinn, B. (2005) Arts festivals and the city. *Urban Studies* 42 (5–6), 927–943. DOI:10.1080/00420980500107250.
Quinn, B. (2006) Problematising 'festival tourism': Arts festivals and sustainable development in Ireland. *Journal of Sustainable Tourism* 14 (3), 288–306. DOI: 10.1080/09669580608669060.
Ramchander, P. (2007) Township tourism – Blessing or blight? The case of Soweto in South Africa. In G. Richards (ed.) *Cultural Tourism: Global and Local Perspectives* (pp. 39–67). New York: Haworth Hospitality Press.
Reid, D. (2002) Development of cultural tourism in Africa: A community based approach. In J. Akama and P. Sterry (eds) *Cultural Tourism in Africa: Strategies for the New Millennium* (pp. 25–34). Conference Proceedings: The ATLAS Africa International Conference. Mombasa: Kenya. Retrieved from http://www.atlas-euro.org/pages/pdf/Cultural%20tourism%20in%20Africa%20Deel%201.pdf.
Reisinger, Y. (2009) *International Tourism: Cultures and Behaviour*. Oxford: Butterworth-Heinemann.
Reisinger, Y. and Turner, L.W. (2012) *Cross-cultural Behaviour in Tourism: Concepts and Analysis*. Oxford: Butterworth-Heinemann.
Richards, G. (2001) *Cultural Attractions and European Tourism*. Wallingford: CABI Publishing.
Richards, G. (2002) Satisfying the cultural tourist: Challenges for the new millennium. In J. Akama and P. Sterry (eds) *Cultural Tourism in Africa: Strategies for the New Millennium* (pp. 35–41). Conference Proceedings: The ATLAS Africa International Conference. Mombasa: Kenya. Retrieved from http://www.atlas-euro.org/pages/pdf/Cultural%20tourism%20in%20Africa%20Deel%201.pdf.
Richards, G. (2007) *Cultural Tourism: Global and Local Perspectives*. Binghampton, NY: Haworth Press.
Smith, M.K. (2009) *Issues in Cultural Tourism Studies* (2nd edn). New York: Routledge.
Snowball, J.D. (2010 *Measuring the Value of Culture: Methods and Examples in Cultural Economics*. Berlin: Springer.
South Africa Tourism (2014) Cultural tourism in South Africa. Retrieved from http://www.southafrica.net/za/en/articles/entry/article-southafrica.net-cultural-tourism-in-south-africa.
Tassiopoulos, D. (2010) *Events Management: A Developmental and Managerial Approach* (3rd edn). Landsdowne: Juta.
Wales Tourist Board (2005) Sense of place toolkit. Cardiff, Whales Tourist Board. Retrieved from http://www.tourisminsights.info/ONLINEPUB/WALES%20TOURIST%20BOARD/WTB%20PDFS/WTB%20%282005%29,%20Sense%20of%20Place%20Toolkit,%20WTB,%20Cardiff.pdf.
Wickens, E. (2005) Cultural heritage tourism – Being, not looking: Beyond the tourism brochure of Greece. In M. Novelli (ed.) *Niche Tourism: Contemporary Issues, Trends and Cases* (pp. 111–120). Oxford: Elsevier Butterworth-Heinemann.
Whithers, G.A. (1980) Unbalanced growth and the demand for performing arts: An econometric analysis. *Southern Economic Journal* 46 (3), 735–742.
Yuan, J.J., Cai, L.A., Morrison, A.M. and Linton, S. (2005) An analysis of wine festival attendees' motivations: A synergy of wine, travel and special events? *Journal of Vacation Marketing* 11 (1), 41–58. DOI:10.1177/1356766705050842.
Zeppel, H. and Hall, C.M. (1992) Arts and heritage tourism. In B. Weiler and C.M. Hall (eds) *Special Interest Tourism* (pp. 47–65). London: Belhaven Press.
Zijlmans, K. and Van Damme, W. (2008) *World Art Studies: Exploring the Concepts and Approaches*. Amsterdam: Valiz.

7 Reflections on International Carnivals as a Destination Recovery Strategy: The Case of Zimbabwe

Cleophas Njerekai

Introduction

International carnivals are creative art forms in which revellers, spectators and participants are treated to aesthetic and dramatic presentations at a mass level (Oluwatoyin, 2011). Through these events, each individual country is afforded an opportunity to celebrate its totality and diversity in terms of its people, food, drink, music, colour, creed, dress and culture among other things. The celebrations involve a public parade, combining elements of circus, masques and a street party (Getz, 2008). Carnivals offer a dynamic tool for self-expression and exploration, and an opportunity to understand and appreciate other cultures. They provide enjoyment and entertainment to millions of people across the globe and hence contribute significantly to the gross national happiness of the countries involved. In relation to tourism development, carnivals play a pivotal role in profiling destinations and provide a platform for managing perceptions. Today, carnivals are celebrated annually in many countries across the globe and they are considered big business in countries such as Brazil, the United Kingdom, Nigeria and many others where they have become a tradition.

Brief History of Carnivals

Existing literature does not provide a clear picture of the history of carnivals. However, there seems to be consensus on the fact that they are not a recent phenomenon and that they originated as a pagan festival in ancient Egypt before the 12th century. They were subsequently celebrated by the Greeks at a spring festival in honour of Dionysus, their God of wine (Rose, 2014). Later on, the Roman Catholic Church in Italy and in other parts of Europe adopted the same concept and called it the festival of 'Carne Vale', thus explaining the derivation of the word carnival – from the Latin words 'carne' and 'vale' meaning 'a farewell to flesh' or 'to put away the meat' (Guitar, 2007). The festival of Carne Vale was a feast celebrated every February starting on Sunday and leading up to just before the Wednesday marking the beginning of Lent. The period of Lent is the 40 weekdays from Ash Wednesday to Easter Sunday, a period of penitence, fasting and fleshly denial in the Christian religion (Getz, 2010). This was all done in recall of the Gospel account of the 40 days and 40 nights Jesus spent fasting in the wilderness.

With time, the carnivals in Italy became quite famous and the practice spread to France, Spain, and to other Catholic countries in Europe. As the French, Spanish and Portuguese began to take control of the Americas and other parts of the world, they also brought with them their tradition of celebrating carnivals. As the carnivals spread across these countries, various celebratory customs were added, making them more and more colourful.

Another explanation of the development of carnivals, by Mauldin *et al.* (2004), recounts that the earliest mention of such a celebration in Europe is recorded in a 12th-century Roman account of the pope and upper-class Roman citizens watching a parade through the city, followed by the killing of steers and other animals. The authors also assert that carnivals began as a Christian festival that was marked with much laughter and excitement and began just before Lent. The purpose was to play and eat as much meat as possible before Ash Wednesday, which marked the beginning of Catholic Lent. The carnival celebration spread with the sweep of Christianity across Europe and eventually into the Americas. According to Getz (2010), although the word 'carnival' originated with this pre-Lenten celebration, the celebratory styles of masking, inversion and grotesquery were also added to these carnivals.

The above developments possibly explain why carnivals are still largely held in Catholic societies, are less common in Eastern Orthodox societies and usually not held in Protestant societies. However, today, carnivals are celebrated around the world for different reasons and not necessarily in the days preceding the onset of Lent.

In Africa, carnivals are not a new concept. It was customary in many places in ancient Africa for people to parade around their villages in traditional regalia, mainly consisting of different types of symbols, feathers and other natural objects. According to Nunley and Bettelheim (1988) these natural objects were meant to bring good luck to the villagers and to scare away evil spirits.

The carnival concept was transported to the Caribbean islands, including Trinidad and Tobago, by the European slave traders. Slaves were mainly gathered from east and central Africa prior to 1834 and the slave masters were excluded from their carnivals. An interesting dimension of the carnivals during this time was the Canboulay, a night-time procession whose original purpose was to gather the slaves together and march them to neighbouring sugar cane plantations to put out fires (Tull, 2005). The burning canes or *cannes brulées* (French) was Canboulay in the local Creole language. The popular Canboulay consisted of a procession with lighted torches (*flambeaux*) accompanied by singing, dancing and drumming (Tull, 2005). The drumming was typically provided by accomplished drummers from the Nigerian Yoruba religion called Shango or Rada (Trinidad) or today Orisha in Trinidad. Tull (2005) goes on to recount that there were also stick fighters armed with three-feet-long bois or staves made from the wood of the Poui. Their fierce Kalenda songs and dances provided a sense of confidence and braveness.

After the ending of the slave trade in 1834, the slave masters abandoned these carnivals and the streets were taken over by the former slaves. Subsequent carnivals became a celebration of the end of slavery and had a new cultural form, derived from the African heritage and the new Creole artistic cultures developed from the Caribbean. This new carnival type included a masquerade that mocked the antics of the former slave masters and acted as a reminder of the evils of slavery (Rose, 2014). As a result, there were many attempts by the now British colonial authorities to suppress and abolish this new type of carnival. These took the form of virulent media campaigns and laws that tried to control the times of the festivals (Nurse, 2001). Licences were required for certain masquerades and the use of drums and flambeaux was banned. The people struggled, fought and died to defend their carnival festival. One famous victory was the defeat of Captain Baker and the special police who were brought to the islands of Trinidad and Tobago from England to suppress the carnival in 1881 (Tull, 2005). These Canboulay riots established the existence and survival of carnivals until the present day. In Trinidad today, the Trinidad Carnival, also known as 'The Greatest Party on Earth', is celebrated every year, in the two days before Ash

Wednesday. It starts in the darkness of the early morning on Sunday with drums, whistles and the beating of iron. People wear masks and soak themselves with mud or oil. Crudely made satirical costumes are portrayed. The start of the carnival is called Jouvay from *Jour Ouvert* (French), meaning daybreak, and is the historical remnant of Canboulay.

In South America, the Africans, Brazilians and Afro-Brazilians during the years of slavery also adopted the carnival culture, which eventually culminated in today's famous Rio Carnival – 'The Greatest Show on Earth', associated with the Samba dance (Rose, 2014).

This brief history shows that carnivals are not a foreign or new concept in Africa, albeit they are in many other parts of the world. It is mainly the dimensions of scale, spread, timing, purpose and staging of carnivals which have evolved over time. As noted by Getz (2010), carnivals have continued to evolve and in places such as Notting Hill, London, Leeds, Yorkshire, etc., they have become divorced from their cycles in the religious year, becoming purely secular events that take place in the summer months (Oluwatoyin, 2011). Many of these international carnivals today have become major standalone tourist attractions.

Background and History of International Carnivals in Zimbabwe

International carnivals are a fairly new phenomenon in Zimbabwe and as a result they have a very short history and there is a dearth of literature on them. Only two international carnivals have been held in Zimbabwe to date – one in 2013 and the other in 2014. The carnival concept was launched in Zimbabwe after the Ministry of Tourism and Hospitality Industry was challenged by the country's vice president in 2011 to start its own carnival through its implementing agent, the Zimbabwe Tourism Authority (ZTA). The vice president raised this challenge after she led a delegation to the Carnaval De Victoria in the Seychelles. The major objective of Zimbabwe's carnival chapter was to provide an international platform for convergence of cultures and a festivity of various arts, which in turn would promote tourism recovery and development in the country. More specific objectives according to the ZTA Harare International Carnivals Cabinet Report (2014) were as follows:

- To increase destination awareness and brand visibility.
- To promote domestic tourism.

- To enhance and nurture artistic and creative talent among locals, particularly the youths.
- To provide a platform for convergence and appreciation of other cultures.
- To motivate participation by locals in tourism development initiatives/programmes.
- To increase tourism revenue.
- To promote local tourism products (souvenirs and food).
- To create domestic tourism awareness through school competitions.
- To promote the music industry through giving them an opportunity to compete in coming up with a theme song for the carnivals.
- To mobilise support and sponsorship from both public and private sectors and highlight the unifying force of carnivals among citizens and nations.

The first carnival in independent Zimbabwe was held in 2013. The carnival attracted close to 1 million spectators from 14 countries. The second edition of the Harare International Carnival attracted 19 countries, among them Brazil, Ethiopia, Italy, Zambia, Trinidad and Tobago, the United Kingdom and the United States (Zhangazha, 2014). The theme for this carnival was 'Celebrating our Diversity'. The week-long celebration of festivities culminated in a street party on 24 May 2014 which witnessed the participation of up to 7000 well-choreographed masquerades, both local and foreign, and an estimated 1.5 million revellers where Trinidad and Tobago was voted the best international group and the overall winner of the carnival (Zhangazha, 2014).

The Zimbabwe International Carnival itself is now a week-long annual festival that encompasses a series of events and festivities aimed at advancing the arts, culture and heritage of Zimbabwe as well as uniting the populace (Problem Masau Arts Correspondent, 2014a). Carnival activities have included the Carnival Queen show followed by a Gospel show, then a Samba night and the Carnival Dance Hall night. Other events have included the Carnival Jazz night, Carnival Exotic night, the Traditional Bira and the Ethiopian night. The climax of the carnival has been the street party and procession starting from the Africa Unity Square and finishing at the Harare Exhibition Park (Problem Masau Arts Correspondent, 2014b).

The Harare International Carnival (HIC) is therefore a unique brand that involves people across all social, cultural and political divides and creates an avenue to promote unity among Zimbabweans, business opportunities and to further strengthen Zimbabwe's mutual relationships with other countries.

Reflections on International Carnivals in Zimbabwe

The international carnivals in Zimbabwe are organised by the ZTA's Festivals, Events and Exhibitions (FEE) Department. This department organised the 2013 and 2014 carnivals and also played a key role in organising the 20th edition of the United Nations World Tourism Organisation (UNWTO) which was jointly hosted by Zimbabwe and Zambia in Victoria Falls in 2013. However, the hosting of the international carnivals presents issues and challenges regarding financing, staging, impacts and others. These need to be reflected on to ensure that this potentially formidable tourism recovery strategy does not degenerate into one of 'those failed me too programmes' for the country.

Carnival planning and financing

The extent of buy-in by the public and other stakeholders was quite limited due to ineffective marketing, as reflected by some citizens who expressed disgust towards the scantily dressed parades from countries such as Brazil. Such reactions are a clear testimony that some sections of society did not fully comprehend and appreciate the concept.

Questions were also raised by the Zimbabwe Chiefs' Council as to why the carnival organisers did not involve traditional leaders who are the custodians of culture if the event had to do with culture. In this regard, interest groups were under-defined or under-represented in the Carnival Organising Committee. There is therefore a need for better community engagement in future carnivals through co-opting the wider community of interests, which includes residents, educationalists and businesses, in the current Carnival Organising Committee.

The 2014 edition of Zimbabwe's International Carnival cost about US$913,000 to organise (Zhangazha, 2014). The chief executive officer (CEO) for the ZTA indicated that the organisation had to rely on corporate sponsors to meet this budget and that the government had only given US$60,000 from the US$200,000 it had promised before the event.

The ZTA CEO indicated that the funds obtained from government were used to host visitors. However, the authority has indicated that it has devised a working plan to cut down the number of hosted visitors in the next five years. He explained that in the next two years, the number of hosted visitors would drop by 50% – the country would meet the costs of half of the invited groups while the remaining half would meet their own costs. The government could not raise the promised amount as the country's treasury was cash-strapped and hence the over-reliance on sponsors. In other countries where international carnivals are popular such as Trinidad and Tobago, the government fully funds these events. For instance, in 2011, South Africa's Cape Town carnival budget was US$1.2 million and the government disbursed US$800,000 (African Economic Outlook, 2014).

The Harare International Carnival is underfunded and the ZTA confirmed that lack of funding would continue to stifle the growth of the carnivals as they mostly relied on sponsors whose contributions could not be guaranteed. The ZTA also confirmed that most of the invited local artists were reluctant to confirm their participation in the 2014 carnival as there was no guarantee of payment (Tlou, 2014). In 2013, when the country launched the carnival, finance was still a challenge and the carnival, which required about US$500,000, was still supported by the private sector. Therefore the country has to come up with new innovative methods to raise the requisite funds for the event.

Carnival Impacts

International carnivals generally have both positive and negative impacts on the host communities and participants (Richards & Palmer, 2010). These impacts can be tangible or intangible and are the *raison d'être* of these events as they shape and mould residents' and other stakeholders' perceptions and attitudes towards the hosting of future carnivals. This section therefore discusses the social-political, economic and environmental impacts of Zimbabwe's international carnivals through the lenses of its set objectives.

Socio-cultural impacts

On the positive side, about 19 countries participated in the 2014 edition of Zimbabwe's international carnival and this created an opportunity for participants to share experiences beyond geographic and historic divides. All of the 19 countries showcased their cultures and the carnival therefore provided a platform for local, regional and international cultural convergence and appreciation as set oout in its objectives. Figure 7.1 shows part of the cultural diversity that has characterised Zimbabwe's international carnivals.

These impacts are in line with what was observed by the UNDP (2010) for Brazil, Colombia, Cuba and Trinidad and Tobago, where carnivals have, to date, provided a concentration of live and recorded music and dance performances that have considerable cultural significance for both domestic and international audiences. To reflect further, the UNDP (2010) observes that

Figure 7.1 Part of the cultural diversity that has characterized Zimbabwe's international carnivals since 2013

such festivals generate cultural value for local people who are afforded an opportunity to project their cultural identities onto the international arena and to enjoy their country's traditional costumes, music, dance and rituals irrespective of their social, physical and ideological differences. However, there is a need to conduct empirical research on this aspect to assess the extent to which carnivals have and can achieve cultural exchange in Zimbabwe and tourist revisit intentions. Existing studies (Bowdin et al., 2006; Jago, 2010; Prayag, 2009) confirm that there is a positive correlation between image, satisfaction and future behaviour, but how overall image influences these constructs remains underexplored (Fredline et al., 2010).

In line with observations by Fredline et al. (2010), the 2014 Harare International Carnival succeeded in motivating local participation in tourism development initiatives through the ZTA-led clean-up campaign. This obviously conjured up a sense of community pride and image enhancement for the places that were cleaned. Some of the organisations that participated in this clean-up included the Zimbabwe Sunshine Group, Clean and Green Miracle missions, Environment Africa, the Environment Management Authority, and Africa Lotto (ZTA, 2014a). During the clean-up, participants were counselled on recyclable materials and how best to turn trash into cash.

In terms of skills transfer, during the second edition of the Harare international Carnival, Danhiko Vocational Training Institution students were trained by participants from Trinidad and Tobago on carnival costume making in a four-day workshop (ZTA, 2014a). The costumes were used for the parade by the masqueraders. Some children with hearing and speech problems were also taught to walk on stilts in two days and they paraded on those at the street party. The Zimbabwe College of Music also facilitated workshops on music production in collaboration with the governments of Brazil and Trinidad and Tobago while Ruff Diamond from the United Kingdom offered a one-day dance workshop focusing on exercises through *soca* music and dance (ZTA, 2014a). This training was also in line with one of the carnival objectives of enhancing and nurturing artistic and creative talent among locals. In addition, the government of Trinidad and Tobago offered ten scholarships at university level in the creative arts at the University of Trinidad and Tobago and the University of the West Indies for studies on event planning and carnival management as part of human capital capacity building for such events. The country's institutions of higher learning could also start offering such programmes. In general, the number of reported crimes and casualty rates for the two carnivals to date has been insignificant.

Zimbabwe's international carnivals held to date have had their share of negative social impacts. First, the ZTA was criticised for cultural shock and indecency as the event was associated with some nudity and erotic dances. Culturally conservative groups cited cultural pollution and commodification as the problem. They also indicated that such dressing was alien to Zimbabwe's culture, with no element of Ubuntu/Hunhu. However, the ZTA defended its position and one official was quoted saying, 'Yes, in our culture it is bad to see people naked, but in this case we were celebrating that cultural diversity.' In addition, another ZTA official said that, 'We do not invite those countries to come and act or behave like us. We need to see their cultures and that is why we invited the Samba girls and others.' These sentiments are also defended by Barker and Hart (2007) who indicate that people who fail to appreciate other cultures are prisoners of their own values, beliefs and norms and suffer from cultural blindness of other cultures.

Other negative impacts of hosting the Harare International Carnival, as cited by the Carnival Organising Committee, were that a number of people were inconvenienced and faced challenges as some of the roads were closed on the carnival day to allow the smooth flow of the carnival procession. On the day of the street party, streets were overcrowded and there were several traffic jams, noise pollution and a lot of litter.

Economic impacts

Not much has been documented about the economic impacts of Zimbabwe's past two carnivals. The ZTA indicated that the 2014 International Carnival was estimated to have drawn at least 800,000 attendees at the start of the carnival rising to a crowd of 1.5 million revellers on the day of the street party on 24 May 2014. This day also witnessed participation of up to 7000 well-choreographed masquerades, both local and foreign. Surveys by the ZTA indicate that each reveller spent an average of US$14, thereby resulting in transactions of about US$21 million. Regrettably, however, the ZTA did not have the capacity in terms of resources and technology to indicate how many domestic, regional and foreign visitors were in Harare specifically for the carnival. The ZTA should therefore come up with mechanisms to measure the economic impacts of these carnivals more accurately. These statistics if positive could provide a compelling case for further and greater investment in future carnivals.

In terms of revenue generation from the other economic activities, a lot of revenue was generated from costume designing, mask manufacturing and trading, etc. Small to medium-sized enterprises (SMEs) were also allowed to create market stalls along the route of the street party and these were maintained throughout the week of the carnival. A lot of cultural artefacts and souvenirs were therefore sold.

A number of musicians and musical groups who were also lined up for the event had revenue and market opportunities created for them. These included Luciano, who performed at the Reggae Night, Brother Valentino, the Brazilian Samba music, Diamond Musica, Bev and Zoey, Oliver Mtukudzi and Trinidad and Tobago's Exodus Steel Orchestra. All in all, now in its second year, the Harare International Carnival has enormous economic potential that has yet to be fully realised.

Environmental impacts

Prior to the 2014 Harare International Carnival, the Harare City Council and the ZTA launched the carnival clean-up programme at Africa Unity Square (Dauramanzi, 2014). The theme for the clean-up campaign was 'Celebrating a cleaner environment'. The clean-up campaign was very significant because the Harare community was encouraged to clean its environment, including their backyards. The community was advised by the country's Environmental Management Authority (EMA) that it was their responsibility to make their environment clean rather than directing blame at the council for failing to collect rubbish (Dauramanzi, 2014). However, as already alluded to earlier, there was a lot of noise pollution and littering on the day of the street party.

Sustainability and the Future of International Carnivals in Zimbabwe

According to Nurse (2001) a festival can create a new tourism season and/or fill a void in the tourism calendar by boosting airlifts and improving hotel occupancies. Zimbabwe's international carnivals are hosted in May which is the low season of the tourism calendar and therefore the carnival has the potential to improve the tourism receipts of the country. Festival tourists are known to stay longer and therefore spend more on local goods and services.

Threats to the sustainability of the country's international festivals lie in the persistence and continuation of deflation, de-industrialisation and the liquidity crunch that the country is currently experiencing. The tendency of the carnival organisers to overpromise and under-deliver may also tarnish the image of future carnivals. All in all, Zimbabwe's International Carnival is poised to play a significant role in boosting tourism in Zimbabwe. This will have spillover effects in the form of promoting peace and social goodwill among the locals and achieving the perceptual shift that the country certainly needs.

International Carnivals and Destination Recovery

Foreign carnival participants and visitors have the potential of marketing destination Zimbabwe through word of mouth. There is also a possibility that these participants may revisit Zimbabwe with families and friends, for exploration and enjoyment of forthcoming carnivals. This will result in increased visitor numbers to destination Zimbabwe. The hosting of such mega events may also send the right signals about Zimbabwe to the rest of the world. The fact that 19 countries and 1.5 million spectators participated in a well-choreographed street parade is in itself a powerful marketing tool for the country. Experiences from other countries such as Kenya have proven that international carnivals have the capacity to enliven destinations. As reported by Jago (2010), the Mombasa International Carnival successfully brought the town of Mombasa back to its former glory after years of municipal neglect.

Therefore, international carnivals have the potential to animate cities and increase a destination's awareness and brand visibility throughout the world market. They can promote domestic tourism and motivate local participation in tourism development initiatives. They can also increase tourism revenue and promote local tourism products.

Conclusion

The future of carnivals in Zimbabwe looks bright. These events have the potential to significantly turn around the country's fortunes in terms of boosting tourist arrivals, generating foreign currency, forging partnerships and external linkages, creating employment and increasing cultural understanding and awareness of the locals. However, the success of the country's previous carnivals hinges on its ability to improve community engagement and to ensure that local Zimbabweans understand the concept and share the vision. This could be done by ensuring that the event becomes a part of every Zimbabwean's way of life – making it a lifestyle as opposed to an event. These is also a need to source more funding and to continuously offer new dimensions to the event so that it remains exciting and relevant to all stakeholders. Other than this, the Harare International Carnival is expected to gain prominence in the near future and is likely to be used as a template for further carnival developments in southern Africa. If leveraged well, these carnivals have the potential to boost tourism in the country. Efforts should therefore be made to ensure that these carnivals develop into truly independent and self-sustaining events. As of now, the real challenge is to conceive of the carnival in new ways, retaining the best of the past and building on its strengths.

References

African Economic Outlook (2014) Tourism for tomorrow. See www.ariseafrica.com/ (accessed 20 August 2014).
All Africa (2014) Zimbabwe: Harare International Carnival on the cards. All Africa. See http://allafrica.com/stories/201302120277.html/ (accessed 21 August 2014).
Barker, M. and Hart, S. (2007) *The Marketing Book*. New York: Butterworth-Heinemann.
Bowdin, G., Allen, J., O'Toole, W., Harris, R. and McDonnell, I. (2006) *Events Management*. (2nd edn). Oxford: Butterworth-Heinemann.
Dauramanzi, F. (2014) Harare International Carnival looking green. *Harare News*, 17 April. See http://www.hararenews.co.zw/2014/04/Harare-international-carnival-looking-green/ (accessed 31August 2014).
Fredline, L., Jago, L. and Deery, M. (2010) *The Development of a Generic Scale to Measure the Social Impacts of Events*. New York: Routledge.

Getz, D. (2008) Event tourism: Definition, evolution and research. *Journal of Tourism Management* 29, 403–428.
Getz, D. (2010) The nature and scope of festival studies. *International Journal of Event Management Research* 5 (1), 1–47.
Guitar, L. (2007) *The Origins of Carnival – And the Special Traditions of Dominican Carnaval* [Online]. See www:http://ciee.typepad.com/files/carnaval-origins_w-photos-1.pdf (accessed 28 August 2014).
Jago, L. (2010) Optimising the potential of mega events: An overview. *International Journal of Event and Festival Management* 1 (3), 223–224. See http://dx.doi.org/10.1108/17852951011078023
Mauldin, B., Walkup, N. and Gomez, A. (2004) *Carnival*. New York: Crystal Productions Company.
Nunley, J.W. and Bettelheim, J. (1988) *Building Bridges Through Culture*. Washington: University of Washington Press.
Nurse, K. (2001) Festival tourism in the Caribbean: An economic impact assessment. See www.acpculture.eu/...//ADB-Nurse-Festival/Tourism (accessed 30 August 2014).
Oluwatoyin, I.S.Y. (2011) Tourism development, poverty and unemployment. See www.ajol.info/journals/jorind (accessed 30 July 2014).
Prayag, G. (2009) Tourists evaluations of destination image, satisfaction, and future behavioural intentions – The case of Mauritius. *Journal of Travel and Tourism Marketing* 26 (8), 836–853.
Problem Masau Arts Correspondent (2014a) Harare Carnival goes national. *The Herald*, 17 April. See www.herald.co.zw/harare-carnival-goes-national/ (accessed 13 August 2014).
Problem Masau Arts Correspondent (2014b) Curtain comes down on carnival. *The Herald*, 23 May. See www.herald.co.zw/curtain-comes-down-on-carnival/ (accessed 13 August 2014).
Richards, G. and Palmer, R. (2010) *Eventful Cities – Cultural Management and Urban Revitalisation*. Oxford: Butterworth – Heinemann.
Rose, M.L. (2014) Carnival origins, carnival in education. See www.carnivalorigion/carnivalin/ (accessed 18 August 2014).
Tlou, P. (2014) Matebeleland artistes to attend international carnival. *The Sunday News*, 11 May. See www.sundaynews.co.zw/mat-artistes-to-attend-international-carnival/education/ (accessed 29 August 2014).
Tull, J. (2005) Money matters – Trinidad and Tobago Carnival 2005. Carnival institute of Trinidad and Tobago. See https://academia.edu/326827/Money-matters in_the_trinidad carnival/ (accessed 31 August 2014).
UNDP (2010) *Creative Economy Report: A Feasible Development Option*. New York: UNDP.
Zhangazha, W. (2014) Still a long way to go for Harare Carnival. *The Zimbabwe Independent*. See www.theindependent.co.zw/2014/05/.../still-long-way-go-harare-carnival/ (accessed 28 August 2014).
Zimbabwe Carnival. See www.http://zimbawecarnival.com/ (accessed 30 August 2014).
Zimbabwe Tourism Authority (ZTA) (2014a) *The Harare International Carnival Cabinet Report*. Harare: ZTA.

Part 2
Impacts and Management of Cultural Tourism

8 The Commodification of World Heritage Sites: The Case Study of Tsodilo Hills in Botswana

Joseph E. Mbaiwa

Introduction

This chapter analyses the effects of resource commodification of World Heritage Sites on local livelihoods and conservation using Tsodilo Hills in Botswana as a case study. Tsodilo Hills was listed as a World Heritage Site (WHS) in 2001. WHS status attracts attention from tourism players, particularly tour operators, tourism developers and tourists themselves (Borges et al., 2011). The label 'Outstanding Universal Value' which often describes a WHS gives tourists the expectation that visiting the site will be a unique experience and at the same time provides the tourism industry with an easily promoted and almost fail-proof destination (Borges et al., 2011). As a result, WHSs are amongst the most popular and heavily promoted attractions in many countries (Buckley, 2004). WHSs are natural or cultural areas designated through the United Nations Educational, Scientific and Cultural Organisation (UNESCO) for protection and preservation for future generations because of their global significance or importance. There are currently 962 WHSs located in 157 countries. Of these, 745 are cultural, 188 are natural and 29 are mixed properties (UNESCO, 2013). Africa has fewer sites listed as WHSs. In Botswana, Tsodilo Hills was inscribed as a WHS in 2001. The Okavango Delta was inscribed as the country's second WHS in June 2014.

There are many stakeholders in WHSs, including local communities who are the traditional owners and custodians of these places. Local communities have very strong cultural connections to WHSs (Borrini-Feyerabend *et al.*, 2004). There are also a multitude of other stakeholders who may also have an interest in the sites. These include local, national and regional authorities, tourists, the tourism industry and the international community (Borrini-Feyerabend *et al.*, 2004). In Botswana, research has not fully established stakeholders in WHSs and how tourism development impacts sites such as Tsodilo Hills, including the nature and extent of different types of effects from tourism in and around sites. As a result, research should establish how different players, such as the tourism industry, tourists, and local authorities, contribute to such effects and which of these groups can be considered the drivers of change in WHSs. In this regard, the commodification of such sites for the tourism market raises questions of conservation and sustainable use due to the increase in tourism activities/development in the sites. The objective of this chapter is to analyse the effects of resource commodification of WHSs on local livelihoods and conservation using Tsodilo Hills in Botswana as a case study. The chapter is, conceptually, informed by the concept of cultural commodification.

Cultural Commodification and Sustainable Tourism

Cultural commodification

The inscription of a site as a WHS often results in the commodification of cultural and natural resources in that particular site for the tourism market. It is critical to understand the effects of commodification of cultural heritage sites to local livelihoods and to the conservation of resources. This chapter used the concept commodification to analyse how tourism development on cultural heritage sites such Tsodilo Hills impacts on local livelihoods and conservation. The concept of commodification has been used by different tourism scholars to analyse cultural tourism development (Ateljevic & Doorne, 2003; Cohen, 1988, 1989; MacCannell, 1973; Steiner & Reisinger, 2006). Commodification is defined as a process by which items and activities are evaluated primarily in terms of their exchange value, in the context of trade, thereby becoming goods (and services); the exchange value of items and activities is stated in terms of market prices (Cohen, 1988: 380). In tourism, the packaging of cultural activities and artefacts for the tourist market is known as the commodification of culture (Cohen, 1988; MacCannell, 1973). In this chapter, commodification of cultural resources refers to the packaging of Tsodilo Hills as a tourism product to be enjoyed by tourists who visit the WHS.

The commodification of culture for the tourism market can have both positive and negative impacts on culture. MacCannell (1973) argues that the commodification of culture for touristic purposes can lead to culture losing its meaning for locals. Cultural commodification changes the meaning of cultural products and human relations, making them eventually meaningless (Cohen, 1988; MacCannell, 1973). Local culture becomes altered and often destroyed by its treatment as a tourist attraction (Cohen, 1988). This means that when local cultures are produced and packaged for the tourist market, their intrinsic value as part of the local cultural identity is lost. When the true authentic culture is lost, what is presented to tourists therefore might not be a true representation of local culture. This means the commodification of local culture creates what is known as staged authenticity (MacCannell, 1973). Staged authenticity undermines genuineness or authentic cultural products since it succumbs to tourists' tastes which may not necessarily reflect the authentic culture (Cohen, 1988). There is therefore little research that has been carried out to demonstrate whether the commodification of Tsodilo Hills as a tourist product has resulted in any loss of local culture, an objective which this study will partly address.

Positively, the commodification of a WHS can increase a community's ability to generate income and improve its well-being. That is, the commodification of culture can also transform non-market cultural aspects into touristic goods thereby adding value to such aspects (Cohen, 1988). Steiner and Reisinger (2006) maintain that tourism development may enhance both the cultural identity and the well-being of members of a local culture. Cultural commodification thus makes communities embrace and preserve those traditions that would have otherwise have disappeared (Cohen, 1988). The commodification of culture can have positive cultural aspects in a community which promotes tourism development as another form of livelihood. This chapter, therefore, should establish the positive impacts of tourism on rural livelihoods due to the commodification of Tsodilo Hills. The expectation in WHSs is that tourism development should be sustainable. Borges et al. (2011) argues that if tourism is badly planned and not managed responsibly, it can on the contrary lead to biodiversity loss, ecosystem degradation and negative impacts to local communities. Therefore, tourism in WHSs needs to be managed such that it upholds the principles of sustainable development. This study, therefore analyses the impacts commodifying Tsodilo Hills WHS on rural livelihoods and conservation.

Box 8.1 Authenticity of the Basotho hat

Haretsebe Manwa

Authenticity of cultural crafts and artefacts has been at the centre of tourism debate for decades (Belhassen *et al.*, 2008; Chang *et al.*, 2012; Cohen, 1993, 2002; Littrell *et al.*, 1993; Maruyama *et al.*, 2008; Olsen, 2002; Roland, 2010; Swanson & Timothy, 2012; Xie, 2011).

Tourists visit places with the hope of gaining authentic experiences (Wang, 1999). This is achieved not only through being true to self – self-discovery – but also seeing and encountering genuine products which are purchased as souvenirs at places visited (Cohen, 1993). As noted by Rogerson, in Chapter 3, selling souvenirs is one of the main economic activities of small towns in southern Africa, for example, regions such as Maseru in Lesotho where the Basotho hat is one of the main products sold in the streets (Rogerson & Sithole, 2001). Its authenticity as a souvenir is of paramount importance.

It has been suggested that the manufacturing process and final product should be used as ways of determining the authenticity of the crafts/artefacts (Cohen, 1992). Cohen proposes that a product that follows the traditional manufacturing methods and where the final product resembles the traditional product in terms of uniqueness, originality, workmanship, aesthetics, function and use, cultural and historical integrity, craftsperson and materials, shopping experience and genuineness would qualify as an authentic product (Littrell *et al.*, 1993).

The Basotho hat (Plate 8.1) is a conical-shaped straw hat made from an indigenous grass – *'mosea'* (Ross, 1976). The Basotho hat is considered part of Lesotho's national dress and is also seen on the national flag and currency. Lesotho is a country in which tradition is still very important, and the wearing of the distinctive Basotho hat is still part of everyday life (Lesotho Government, 2006). The Basotho hat is also associated with King Moshoeshoe I the founder of the Basotho nation.

The Basotho hat fulfills the typologies outlined by Cohen (1992) and Littrell *et al.* (1993). The manufacturing process still follows the traditional ways in that the raw material, the *'Leholi'* or *'Mosea'*, which is the grass used for weaving the hat, is sourced from companies that buy it from harvesters in the highlands of Lesotho who then transport it to various outlets in the lowlands of Lesotho. Elders have transferred the art of weaving the hat from generation to generation. To be able to weave a hat ready for the market, a weaver must have gone through an apprenticeship, in which the mentor certifies that the protégé has reached a stage whereby they can weave without supervision. Also, the final hat reflects similar shapes and designs which have stood the test of time (see Plate 8.1). The only innovation is the inscription of 'Lesotho' to identify the country of origin.

In conclusion, commodification of the Basotho hat has been instrumental in empowering communities, alleviating poverty and enhancing cultural awareness in Lesotho and beyond (Scheyvens, 2003). Cultural artefacts like the people's culture are dynamic (Wherry, 2006). There have been cosmetic changes to the hat. It is not unusual for cultural objects to change over time. Similar observations have been made in Botswana for example, where traditional baskets have changed in shape and design to represent changes in consumer tastes (Mbaiwa, 2004). Most southern African countries have similar artefacts sold by street vendors such as soapstone versions of the big five animals sold in Zimbabwe, earrings, and other artefacts depicting traditional products. Authenticity of these products should be established as a marketing tool for such products and to guard against imitation products which are

Plate 8.1 The Basotho hat
Source: Haretsebe Manwa's photo, 2014.

(continued)

Box 8.1 Authenticity of the Basotho Hat *(continued)*

likely to flood local markets, thereby depriving the informal sector traders of their livelihoods.

References

Belhassen, V., Caton, K. and Stewart, W.P. (2008) The search for authenticity in the pilgrim experience. *Annals of Tourism Research* 35 (3), 668–689.

Chang, J., Wall, G. and Hung, J.R. (2012) Tourists' perceptions of aboriginal heritage souvenirs. *Asia Pacific Journal of Tourism Research* 17 (6), 684–700.

Cohen, E. (1992) Pilgrimage centres: Concentric and excentric. *Annals of Tourism Research* 19, 33–50.

Cohen, E. (1993) The heterogenization of a tourist art. *Annals of Tourism Research* 20 (1), 138–163.

Cohen, E. (2002) Authenticity, equity and sustainability in tourism. *Journal of Sustainable Tourism* 10 (4), 267–276.

Lesotho Government (2006) *Support to Institutional and Capacity Strengthening of the Tourism Sector* (Report on Priority Areas). Maseru: Government of Lesotho; United Nations, Development Programme UN World Tourism Organization.

Littrell, A.M., Anderson, F.L. and Brown, J.P. (1993) What makes a craft souvenir authentic? *Annals of Tourism Research* 20 (1), 197–215.

Maruyama, N.U., Yen, T. and Stronza, A. (2008) Perception of authenticity of tourist art among Native American artists in Santa Fe, New Mexico. *International Journal of Tourism Research* 10, 453–466.

Mbaiwa, J.E. (2004) Prospects of basket production in promoting sustainable rural livelihoods in the Okavango Delta, Botswana. *International Journal of Tourism Research* 6 (4), 221–235.

Olsen, K. (2002) Authenticity as a concept in tourism research: The social organisation of the experience of authenticity. *Tourism Studies* 2 (2), 159–182.

Rogerson, C.M. and Sithole, P.M. (2001) Rural handicraft production in Mpumalanga, South Africa: Organisation, problems and support needs. *South African Geographical Journal* 83 (2), 149–158.

Roland, L.K. (2010) Tourism and the commodification of Cubanidad. *Tourism Studies* 10 (3), 3–18.

Ross, D. (1976) Culture and decorative arts of Africa. *African Arts* 9 (4), 76–77.

Scheyvens, R. (2003) *Tourism for Development, Empowering Communities*. New Jersey: Prentice Hall.

Swanson, K.K. and Timothy, D.J. (2012) Souvenirs: Icon of meaning, commercialization and commoditization. *Tourism Management* 33, 489–499.

Wang, N. (1999) Rethinking authenticity in tourism experience. *Annals of Tourism Research* 26 (2), 349–370.

Wherry, F.F. (2006) The social sources of authenticity in global handicraft markets evidence from northern Thailand. *Journal of Consumer Culture* 6(1), 5–32.

Xie, P.F. (2011) *Authenticating Ethnic Tourism*. Bristol: Channel View Publications.

Study Area – Tsodilo Hills

This study was carried out at Tsodilo Hills, located in north-western Botswana (Figure 8.1). Tsodilo Hills is a WHS whose outstanding universal value is its archaeological account of human activities and environmental changes that span a period of at least 100,000 years. The hills have been Botswana's national monument since 1927. In 2001, the hills were listed by the World Heritage Committee under the cultural criteria (i), (ii) and (vi) as a World Heritage Site (Department of Museum and National Monuments (DMNM), 2005). The registered area of the WHS includes two components: a core zone of approximately 4800 ha and a buffer zone of about 70,400 ha surrounding the core zone. Tsodilo Hills is characterised by the many caves, rock shelters, seasonal camps and a human settlement. There is also much pertaining to the paleo-environment within the site's sediments. This diverse and lengthy chronicle of African cultural heritage and climatic changes provides insights into past human ways of life and how people have interacted

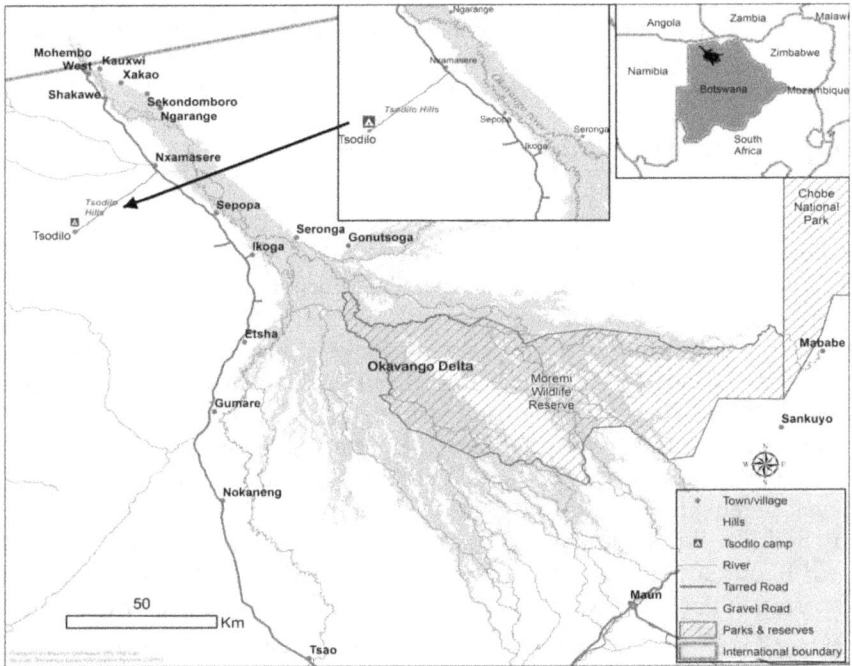

Figure 8.1 Map of Botswana showing the location of Tsodilo Hills and Okavango Delta
Source: Author.

with their environment both through time and space. DMNM (2005), in the Tsodilo Management Plan of 2005, notes that the over 4500 rock art paintings and numerous carvings provide insights into how earlier people perceived their surroundings and artistically expressed their way of life.

At Tsodilo, there is a small village comprising less than 200 people who are from the Basarwa and Hambukusu ethnic groups. The Tsodilo community is involved in a number of livelihood activities, these include: cultural heritage tourism, subsistence cattle farming, crop production and the collection of veld products. Agriculture is the main livelihood activity. Veld products or natural resource harvesting of edible and medicinal plants is still an important part of the community's livelihood with many different species being harvested. Timber is used as fuel wood and for craft and farm/household implement production. Tsodilo Hills as a WHS has become a tourism attraction for both national and international tourists. Visitors to the Tsodilo Hills WHS are diverse in nature, ranging from school children from Botswana and church groups to private self-drive visitors from all over the world. Remnant wildlife populations in Tsodilo Hills exist, the Tsodilo community has thus formed and registered the Tsodilo Community Trust (TCT) in order to benefit from tourism development at Tsodilo Hills. As a WHS, Tsodilo Hills offers the opportunity for tourism.

Data Collection Methods and Analysis

Data for this chapter were derived from both primary and secondary sources. Secondary data sources consisted of articles and reports on Tsodilo Hills as a WHS, community-based tourism projects, and local participation in tourism development and natural resource conservation in the area. This included government policy documents, consultancy reports, the Tsodilo World Heritage Site Dossier, the Integrated Tsodilo Management Plan of 2005, and other natural resource management reports. Information derived from these sources includes historical data on rock paintings, listing of Tsodilo Hills as a World Heritage Site, tourism development in the hills and local participation in tourism development in the hills. Primary data were derived from past and ongoing research field-based surveys at Tsodilo Hills. Some of which has already been reported in documents on tourism development and related environmental conservation issues. The main tool used to collect primary data was face-to-face informal interviews with different interest groups at Tsodilo Hills WHS. These groups include government officials (e.g. the DMNM), non-governmental organisations (NGOs) such as Letloa Trust, donor agencies such as Diamond Trust and local chiefs at Tsodilo Hills. In total, about seven interest or stakeholder groups were interviewed.

Finally, data collected were largely analysed qualitatively. In this regard, thematic analysis was the main approach used to analyse all data collected. Thematic analysis involves the reduction of data into themes and patterns to be reported. Leininger (1985: 60) argues that in thematic analysis themes are identified by 'bringing together components or fragments of ideas or experiences, which often are meaningless when viewed alone'. In thematic analysis, themes that emerge from the informants' stories are pieced together to form a comprehensive picture of their collective experience (Aronson, 1994). In this study, qualitative data from key informant interviews and that collected from secondary sources were summarised into specific themes and patterns on the effects of Tsodilo Hills as a WHS for local livelihoods and conservation.

Results and Discussion

Stakeholders' interests at Tsodilo

Since 2001, when Tsodilo Hills was inscribed as a WHS, there has been growing interest by different stakeholders (Table 8.1) on the cultural heritage site. These stakeholders can be subdivided into groups, namely: government, NGO; private tourism sector, local community and academic research organisations. The involvement of different stakeholders has resulted in Tsodilo Hills becoming one of the key cultural tourism destination sites in Botswana in the last decade.

The involvement of stakeholders at Tsodilo Hills WHS has transformed the site to become one of the key tourism destinations in northern Botswana. These stakeholders include:

(a) Department of Museum and National Monuments

Through an Act of parliament the DMNM is charged with the responsibility of ensuring that all cultural heritage sites in Botswana are preserved and protected. The DMNM has, therefore, produced various planning documents guiding the management and development of Tsodilo Hills through time. The *Tsodilo Hills Management Plan, Scheme for Implementation* was the first management plan produced by DMNM for Tsodilo Hills in 1994 (Campbell, 1994). DMNM also produced and submitted a dossier known as the *Tsodilo, Mountain of the Gods, World Heritage Nomination Dossier* in 2000 detailing various areas of interest in order for Tsodilo Hills to gain world heritage status. This was followed by the *Draft Final Interim Management Plan*

Table 8.1 Stakeholders and stakeholder interest at Tsodilo Hills

Stakeholder	Interest	Comment
Department of Museum and National Monuments	(i) Preservation of San cultural heritage (ii) Maintaining Tsodilo Hills as a WHS	(i) Constructed Tsodilo Hills Museum (ii) Listed Tsodilo Hills as cultural heritage site in 2001 (iii) Production of the Tsodilo Hills Management Plan in 2005 (iv) Assisted local communities in developing cultural tourism projects
Tsodilo Hills community	Deriving of socio-economic benefits from Tsodilo Hills	Formation of Tsodilo Hills Conservation Trust to facilitate cultural tourism projects on behalf of the Tsodilo community
Botswana Tourism Organization	Promotion and development of cultural heritage tourism	Listed Tsodilo Hills as one of the seven key sites to develop cultural tourism
Letloa Trust	Preservation of San culture and improvement of local livelihoods	Work with various organisations in the development of a management plan
Diamond Trust	(i) Promotion of local livelihoods from cultural heritage tourism (ii) Conservation of Tsodilo as a WHS	Provided BWP10 million (US$1.25 million) to Tsodilo Community Trust as seed funds for various community tourism projects
University of Botswana and other institutions	Academic research on cultural heritage tourism	Various researchers have published academic journal articles, books, theses, dissertations and magazine publications about Tsodilo Hills as a cultural heritage site
Tourists	Enjoy the cultural heritage products offered at Tsodilo Hills	There is a relative increase in international and local tourists visiting Tsodilo Hills annually

(Ectocon, 2003), which has been finalised as the Tsodilo World Heritage Management Plan (TWHMP). In 2005 the DMNM finally produced the Tsodilo Integrated Management Plan.

The DMNM has built the Tsodilo Hills Museum at Tsodilo Hills. The museum is managed by DMNM staff and has increased the cultural tourist attraction and interpretation of Tsodilo Hills. The department further mobilises and provides cultural heritage training for guides who take tourists on excursions and tours of the hills. This shows that over the last decade the DMNM has become one of the major interest groups in the protection of Tsodilo Hills, both as a cultural heritage tourism destination and as a cultural heritage site.

(b) Diamond Trust – Donor funding

The Diamond Trust has so far provided the largest amount of funding to the TCT to develop cultural heritage tourism sites. Diamond Trust is a 50-50 corporate social investment venture between Debswana Diamond Company and De Beers Botswana Holdings. In July 2013, Diamond Trust sponsored TCT with an amount of BWP10 million (US$1.25 million). This money was to be used to conserve the site and its surrounding environment through the development of ecotourism projects by the people of Tsodilo Village, particularly campsites and curio shops. Diamond Trust's goal is to contribute to the improvement of local people's livelihoods through the upgrading of the tourism services and facilities. As shown in Table 8.2, the sponsorship by the Diamond Trust has so far resulted in a number of developments at Tsodilo Hills.

The Diamond Trust-sponsored projects were officially launched by the Minister of Environment, Wildlife and Tourism on 30 June 2013. The Diamond Trust CEO noted that the success of the Tsodilo project has been made possible by the Tsodilo Management Authority Board, the DMNM, the TCT, Letloa Trust, the North-West District Council and Tawana Land Board. This again demonstrates that funding agencies have a keen interest in

Table 8.2 Projects sponsored by Diamond Trust at Tsodilo Hills

- Training of 14 guides, of which five are of San (Basarwa) origin while nine are of Hambukushu origin. The training was meant to enhance their skills in guiding tourists around the hills.
- Drilling of two boreholes and reticulation of water to the Tsodilo community.
- Construction of two project staff houses. These are staff members charged with the responsibility of ensuring that funded projects are carried out successfully.
- Employment of seven staff members: a project manager, two project officers, a curio shop manager, two campsite staff and one borehole attendant.
- Holding of three annual HIV/AIDS training workshops for the community of Tsodilo, facilitated since 2010.
- Three community campsites around Tsodilo Hills to generate income for the trust for communities.
- A craft centre for the community to sell their crafts.

Tsodilo Hills and the activities of donor agencies have several socio-economic and environmental impacts on this WHS.

(c) The Tsodilo village community

The Tsodilo village community is a key stakeholder which has direct and indirect impacts at Tsodilo Hills. The Tsodilo community comprises approximately 200 people who live in the buffer zone of the WHS. This community is made up of two main cultural groups, namely: the Ju/'hoansi and the Hambukushu (the latter being greater in numbers). The Tsodilo community has since formed the TCT to benefit from cultural tourism development at Tsodilo Hills. A community trust is a registered legal entity, formed in accordance with the laws of Botswana to represent the interests of the communities and to implement their management decisions in natural resources use. That is, trusts are formed by groups of people living in the same area and sharing common interests in order to benefit from the natural resources found around them (DWNP, 1999). In the case of TCT, the institution is made up of all members of Tsodilo village and its aim is to utilise tourism products like Tsodilo Hills to generate jobs and revenues for the benefit of the members of the community.

The TCT collaborates with DMNM, Letloa Trust and Diamond Trust to develop projects that will create employment opportunities, provide capacity building in terms of tourism skills for local people, and develop tourism projects like campsites and curio shops to benefit from the growing tourism industry at Tsodilo Hills. The community and wider cultural ties of different peoples to the Tsodilo Hills is strengthened and the visitor cultural experience deepened. There is long-term training and support for community members and entrepreneurs from supporting NGOs and the museums to enable the community members to assist in the achievement of the vision. The expansion of tourism development at Tsodilo Hills with the participation of the local community thus has several socio-economic and environmental impacts in the area.

(d) Tourists

International and domestic tourists or visitors form one of the main stakeholders at Tsodilo Hills.

The hills attract a very diverse visitor group, from regular tourists (international, regional and local), school groups for education purposes and cultural exposure, to religious groups who have some religious affinity with the hills (DMNM, 2005). Visitors come to the hills because of their cultural attributes, receiving a deeper understanding of their roots and the people of Botswana.

The rock artwork on the hills is the main attraction to the area, and with the hills gaining WHS status this has resulted in an increase in the number of visitors (DMNM, 2005). The hills and the rock art are able to absorb the needs of the visitors. There are different activity opportunities available, people can be guided to sensitive sites, are free to walk the more adventurous trails, and can spend time in the beautiful environment and draw on its spiritual strength. There has been an annual growth rate of 40.5% of the number of tourists visiting the hills for the last three years (2301 visitors in 2001; 5096 in 2003). Informal interviews with members of the DMNM indicate that tourists increased to over 15,000 in 2012. The busiest season runs between the months of July and September each year. Visitors to the hills vary tremendously, i.e. church groups, school groups, tour busses, mobile safaris and self-drive tourists. Variation also occurs with regard to the length of stay of the different types of visitors, from day trips to week-long trips. School groups and church groups have been identified as having the highest direct impact on the hills.

There is a wide selection of accommodation opportunities available in the core visitor area of the WHS. The options for overnight stays vary from subtle upmarket lodges to well-run community campsites. There are wilderness camps also available where people can experience the sense of the Kgalagadi (this is a dry sandy environment, i.e. a semi-arid environment) during their visit. The DMNM perceives its role as protecting the hills from the pressure of visitor demands whilst continually improving the quality of the visitor experience through cost-effective management, visitor education and support from the neighbouring community. They also see themselves as the focus of the area's sustainable economy by being the watchdog of standards and manager of the source. The DMNM has changed its infrastructure to cater for temporary visitors and to control entry by charging a modest fee.

The increasing number of tourists at Tsodilo Hills WHS has resulted in the development of tourist infrastructure within the core zone. Such infrastructure includes telecommunication facilities and generators. There is also an airstrip, small museum, and museum office; an information centre, ablution block, camping facilities and staff accommodation have been developed within the core zone. The Tsodilo Hills can be accessed via either road or air. Three roads service the area, two of which require 4x4, and a single all-weather gravel road exists. The development of this gravel road has resulted in an increase in traffic and a change in visitor type to the hills as access is now easier (Ecosurv, 2005).

Impacts of tourism at Tsodilo Hills WHS

The commodification of Tsodilo Hills for the tourist market has several benefits that accrue to different stakeholders; Table 8.3 provides a summary.

Table 8.3 Impacts and benefits to stakeholders

Stakeholder	Benefits/impacts
Department of Museum and National Monuments	(i) Promotion of cultural heritage information through Tsodilo Hills Museum, and rock paintings in the hills (ii) Protection of the cultural heritage site of Tsodilo Hills
Tsodilo Hills community	(i) Tourism benefits such as employment opportunities, revenue generation, tourism infrastructure development (e.g. roads, airstrip, campsites, curio shop etc.) (ii) Development of entrepreneurship skills, e.g. professional guiding, marketing, business management, etc.
Botswana Tourism Organization	Has so far not embarked on major tourism projects at Tsodilo Hills
Letloa Trust	Achieving its goal of facilitation of community projects that promote improved rural livelihoods and conservation
Diamond Trust	Achieving its goal of social responsibility through funding of community tourism activities at Tsodilo Hills Community
University of Botswana and other institutions	Publication of cultural heritage information about Tsodilo Hills in academic journal articles, books, theses, dissertations and magazine publications
Tourists	(i) Gaining of knowledge about cultural heritage of San communities of Tsodilo Hills (ii) Relaxation in a cultural heritage and nature-based tourism environment

The introduction of tourism has resulted in the commodification of San culture at Tsodilo Hills WHS. Commodification of culture has positive and negative effects on people's cultural practices. On the positive aspect, the people of Tsodilo Hills are now able to tell stories about their way of life, and perform music and dance and other ritual activities that include traditional hunting demonstrations and bush walks to tourists. In this regard, there has been a revival and cultural preservation of San communities. Informal interviews with key informants in the village of Tsodilo indicate that apart from the income they derive from cultural heritage tourism at Tsodilo Hills, they appreciate tourism in the area because it encourages the preservation of their cultural values and practices. In addition, cultural heritage tourism at Tsodilo Hills has made the local culture known worldwide and enabled it to be researched so that information can be preserved in books for future generations.

On the negative side, interviews with key informants indicate that commodification of Tsodilo Hills for the tourism market has led to restrictions in hunting, especially big game hunting. Informal interviews with key informants indicate that hunting has been the backbone of San culture; hence, its restriction suggests that San children may never know how to hunt like they used to do. Despite these restrictions, key informants seem to agree that cultural tourism has the potential to improve livelihoods in their village.

The commodification of Tsodilo Hills for the tourism market has positive benefits. For example, it has increased the Tsodilo community's ability to generate income and improve its well-being. It has also transformed non-market cultural aspects into touristic goods, thereby adding value to such aspects. For example, cultural items such as baskets, beads and wood cravings are now made by the Tsodilo community and sold to tourists at the craft shop. In this regard, the commodification of Tsodilo Hills is enhancing the cultural identity and the well-being of members of a local culture. Cohen (1988) notes that, in the long term, communities manage to embrace and preserve those traditions that would have otherwise disappeared. The commodification of Tsodilo Hills creates positive tourism and contributes to the rural local livelihoods of local people.

The formation of the TCT to advance the objectives of improved livelihoods, poverty alleviation and conservation of natural resources through cultural heritage tourism is a good gesture in achieving sustainable tourism. Presently, though minimal, the trust has generated socio-economic benefits such as the creation of employment and business opportunities for the community. There are opportunities for sustainable tourism and the improvement of livelihoods through the commodification of Tsodilo Hills. However, there are challenges in that cultural heritage tourism is a new economic sector at Tsodilo village and in Botswana as a whole. In this regard, cultural heritage tourism is not well enough understood, especially by the local communities, for them to significantly decide on the best ways to successfully derive benefits from it. In addition, the community lacks the necessary marketing, entrepreneurship and managerial skills to meaningfully benefit from tourism development in their local environment. DMNM (2005) notes that guiding activities at Tsodilo Hills WHS are of poor quality and few of the guides have formal training.

A study of potential benefits at Victoria Falls in Zimbabwe, which is also a WHS has shown that when a site is designated a WHS (at least in the African context), there are potential socio-economic and environmental benefits associated with it. Table 8.4 shows some of these benefits, which in essence also apply to Tsodilo Hills.

Tourism at Tsodilo Hills WHS has its challenges. The different stakeholders compete and conflict with each other. For example, during a workshop

Table 8.4 Potential benefits at Tsodilo Hills

Enhance economic opportunity
- Increase job opportunities for local residents – employment of the local community in tourism establishments (camps and related enterprises)
- Increase in income – supply of goods and services to tourism enterprises by the local community. Offering more locally made goods for sale to visitors helps increase visitor expenditure and local incomes
- Direct sales of goods and services to visitors by the local community (informal economy)
- Establishment and running of tourism enterprises by the local community – e.g. micro, small and medium-sized enterprises, or community-based enterprises (formal economy)
- Generation of local tax revenues – enhance taxes and levies on tourism income or profits with proceeds directly benefiting the local community
- Improves local facilities, transportation and communications
- Investment in infrastructure (roads) stimulated by tourism which will in turn benefit the local community

Protection of natural and cultural value
- Development of good environmental practices and management systems to influence environmentally friendly tourism businesses
- Communicate and interpret conservation principles to visitors and local community through education. This will encourage local people to value their local culture and environments
- Promotion of aesthetic, spiritual and other values related to Tsodilo Hills

Source: Modified from African Wildlife Fund (2010).

held in Shakawe in June 2013, the village leadership reported displeasure and lack of confidence in how government departments, and non-governmental and parastatal organisations are facilitating cultural heritage tourism in the area. The village leadership perceived government to have fenced the core zone of the WHS, hence depriving them of access to some of their traditional livelihoods such as watering their livestock from spring water from the hills, subsistence hunting and the gathering of veld products. These assumptions were made based on the fact that the long-promised alternative borehole located far away from the core zone was not functional. In this regard, the traditional leadership felt their traditional livelihoods system had been infringed upon by new developments in cultural heritage tourism. In addition, the San and Hambukushu communities transferred limited social capital to tourism development in the area. For example, there are two traditional leaders at Tsodilo Village; the San or Basarwa have their own leader while the

Hambukushu have their own. The lack of cohesion between these groups presents a challenge for the successful implementation of cultural heritage tourism at Tsodilo Hills by the community. There is also an increased risk of the spread of diseases such as HIV/AIDS as a result of the increase in development, trade and general exposure to the outside world (DMNM, 2005).

Environmental impacts are also expected due to the commodification of Tsodilo Hills. DMNM (2005) divides visitor impact into three categories: high impact, medium impact and low impact. High impact visitors are those such as school groups and church groups who visit the hills in large numbers and need to be controlled. Groups of school children require the most input due to their volume and nature. Impacts from school groups range from the killing of wildlife (tortoises) to the carving of names into trees. Church groups are considered the next most intensive group simply due to their volume and the extraction of water from the hills as a result of spiritual beliefs (the Chokam water level is usually reduced after church groups have visited the area). Medium impact visitors are those groups that are large but require less management, i.e. overlanders, tour buses and guided groups. Low impact visitors are small groups or individuals that require little in the way of management: mobile safaris and private individuals. Table 8.5 provides a summary of the environmental impacts at Tsodilo Hills WHS.

There are nine campsites at Tsodilo under the control of the DMNM (2005). Only the campsite near the museum has facilities, and it is usually used for official visits by schools and other institutions. The major threat to the heritage site is that there is currently no limit to the number of people allowed to camp at each campsite. Previous records indicate that large church

Table 8.5 Summary of environmental threats at Tsodilo Hills

- Large groups of school children are difficult to control and invariably end up touching paintings.
- Large religious groups tend to make use of the water found in the hills in the various 'springs', and depletion of these resources is problematic.
- Uncontrolled tourism due to no access control by the museum.
- Uncontrolled research can result in vandalism due to uncontrolled excavation.
- Off-road driving by researchers, tourists, government workers and other visitors.
- Bush fires caused by uncontrolled camping.
- Poor guiding levels.
- Unlimited accommodation facilities create competition for any development of a community-run campsite.

Source: DMNM (2005).

groups, of over 400 people, have been known to occupy a single site (Ecosurv, 2005). Finally, Ecosurv (2005) assumes that the largest impact on the area is uncontrolled access to the hills, resulting in vandalism or unintentional damage to the paintings, thus causing loss of panels. Tourist conflicts (due to 'overcrowding') are possibly the next major impact affecting the area (DMNM, 2005).

There are also impacts reported on the livelihoods of the people of Tsodilo Hills and wildlife populations in the area. For example, DMNM (2005) notes that subsistence harvesting of veld foods, medicinal plants and building materials is a large part of the community's income, with most harvesting occurring within a short radius of the village due to the restriction of access to the different areas. Wildlife populations in the buffer area are also minimal due to various factors, namely human pressures (poaching/ illegal harvesting and livestock competition) and the lack of permanent surface water; also, the erection of a livestock disease veterinary fence, such as that at Ikoga near Okavango Delta, prevents previous localised migrations to the hills from occurring. As a result, wildlife is presently restricted to a few small locations scattered in the buffer zone. DMNM (2005) argues that the population increases are a result of seasonal influxes of some plains game and a few elephant to the area. Leopards are the main predator reported in the area (reported by the community of killing goats), whilst species such as lion have not been seen or heard in the area for a considerable amount of time (DMNM, 2005). This shows that, despite the fact that Tsodilo Hills is now a WHS, there are impacts associated with the fencing of the area in the attempt to create a core and buffer zone for the site.

Conclusion

World heritage properties like Tsodilo Hills are important travel destinations which have impacts on livelihoods and conservation. However, for Tsodilo Hills WHS there is a mixed picture, which includes stakeholder conflicts, a relatively low impact on rural livelihoods, and environmental threats to the WHS due to increased visitor numbers. If tourism development occurs rapidly and without planning or appropriate regulatory control, it makes a WHS vulnerable to over-utilisation and environmental degradation. For example, the increase and upward trend in visitor numbers at Tsodilo Hills has resulted in vandalism of rock art and environmental impacts caused by the collection of water by spiritual tourists (DMNM, 2005). However, DMNM developed an Integrated Management Plan for Tsodilo Hills in 2005 in order to sustainably manage human activities such as tourism development. It is assumed that this will ensure tourism has a positive impact on local livelihoods and will achieve environmental sustainability.

In promoting the environmental conservation of Tsodilo Hills as a WHS, the Tsodilo Management Plan argues that tourism and community development, particularly local livelihoods, should be integral in the management of the area, and that the spiritual connection of communities to the hills needs to be recognised. Tourism development should also focus on developing an internationally recognised, comprehensive and quality visitor experience through the provision of archaeological, cultural and wildlife tourism components. As a result, a diversified tourism resource base should be made to reduce pressure on the existing archaeological sites, thus protecting the biodiversity resources of the core area and managing the buffer area in a sustainable manner.

References

African Wildlife Fund (AWF) (2010) World heritage sites and sustainable tourism – Situational analysis: Victoria Falls World heritage site. Unpublished report, African Wildlife Fund.
Aronson, J. (1994) A pragmatic view of thematic analysis. *The Qualitative Report* 2, 1–3.
Ateljevic, I. and Doorne, S. (2003) Culture, economy and tourism commodities: Social relations of production and consumption. *Tourism Studies* 3, 123–141.
Borges, M.A., Carbone, G., Bushell, R. and Jaeger, T. (2011) *Sustainable Tourism and Natural World Heritage – Priorities for Action*. Gland, Switzerland: IUCN.
Borrini-Feyerabend, G., Kothari, A. and Oviedo, G. (2004) Indigenous and local communities and protected areas: Towards equity, and enhanced conservation. Best Practice Protected Areas Guideline Series No. 11. Gland: IUCN.
Buckley, R. (2004) The effects of World Heritage listing on tourism to Australian National Parks. *Journal of Sustainable Tourism* 12 (1), 70–84.
Campbell, A. (1994) *Tsodilo Hills Management Plan, Scheme for Implementation*. Gaborone, Botswana: Department of National Museum and Monuments.
Cohen, E. (1988) Authenticity and commoditization in tourism. *Annals of Tourism Research* 15, 371–386.
Cohen, E. (1989) 'Primitive and remote': Hill tribe trekking in Thailand. *Annals of Tourism Research* 16, 30–61.
Department of Museum and National Monuments (DMNM) (2005) *Tsodilo Hills World Heritage Site Integrated Management Plan*. Gaborone, Botswana: Department of National Museum and National Monuments.
Department of Wildlife and National Parks (DWNP) (1999) *Joint Venture Guidelines*. Gaborone: Department of Wildlife and National Parks.
Ecosurv (2005) *Tsodilo Hills World Heritage Site Integrated Management Plan*. Gaborone, Botswana: Department of National Museum and National Monuments.
Ectocon (2003) *Tsodilo Interim Management Plan-Draft Final: 2004*. Gaborone, Botswna.
Leininger, M.M. (1985) Ethnography and ethno-nursing: Models and modes of qualitative data analysis. In M.M. Leininger (ed.) *Qualitative Research in Nursing* (pp. 33–72). Orlando, FL: Grune & Stratton.

MacCannell, D. (1973) Staged authenticity: Arrangements of social space in tourist settings. *American Journal of Sociology* 79, 589–603.

Steiner, C.J. and Reisinger, Y. (2006) Reconceptualising object authenticity. *Annals of Tourism Research* 33, 65–86.

United Nations Educational, Scientific, and Cultural Organization (UNESCO) (2013) UNESCO World Heritage Sites Lists - 2013. United Nations Educational, Scientific, and Cultural Organization. Unpublished Report. Paris: World Heritage Centre.

9 Tourism and the Social Construction of Otherness through Traditional Music and Dance in Zimbabwe

Patrick Walter Mamimine and
Enes Madzikatire

> *Undeniably, global tourism is the quintessential business of difference projection and the interpretive vehicle of Othering par excellence, with many peoples now cleverly Othering themselves.*
> Salazar (2013: 690)

Introduction

Cultural tourism thrives on the visitor's desire to experience 'Otherness'. Nevertheless, modernisation and globalisation have seriously undermined any society's claim to absolute cultural distinctiveness. In these circumstances the authenticity of the experience of 'Otherness' becomes a contested reality especially where aspects of a society's culture are showcased on the stage (Kroflic, 2007; Lichem, 2012). The challenge then for most destinations offering culture as a tourism product outside their natural setting is to allay the visitors' fears of being exposed to some form of inauthentic Otherness.

There is significant convergence in the various scholars' conceptualisation of Otherness in the literature. The term is construed as 'applying a principle that allows individuals to be classified into two hierarchical groups: them and us' (Staszak, 2008), people who look too different or behave too differently, or who see the world too differently (Blatt, 1987), absolute difference (Levinas, 1987), a quality of being not alike; being distinct or different from that which is otherwise experienced or known (Mengstie, 2011) and some form of difference made relevant depending on context (Soyland, 2006). The major difference among scholars lies in their views over the continuum of difference defining 'Otherness'.

Mengstie (2011) observes that the experience of being 'other' can be expressed in many ways. This chapter is a critical narrative of the social construction of Otherness through the Shona people's traditional music and dance performed for a tourist audience in Harare (Zimbabwe) at Chapungu Sculpture Park. The chapter presents part of fieldwork findings of a study conducted over 18 months for a doctoral degree of the first author.

Data were collected through observations and in-depth interviews held with the dance group leader and tourists. Interviews with tourists complemented observations in order to gain more insight on the impact of the traditional dance and music portrayed on the visitors' expectation of the experience of Otherness. The study employed thematic analysis of the data to make sense of the phenomena under investigation.

Among the Shona people of Zimbabwe, music and dance play an important role in celebrations and marking life-cycle events such as death, marriage and many others (see Chiwome, 1996). Normally, there is no audience at such events. Everybody is a participant in any of the roles of drumming, clapping hands, playing rattles, dancing, singing and whistling to urge on those on the dancefloor. In such a scenario, the authenticity of both the occasion and the experience is uncontestable.

However, the expression of Otherness through traditional music and dance for tourism recreation undergoes what Dietvorst (1994) refers to as material and symbolic reproduction by both producers and consumers. In material and symbolic reproduction the producers interfere with the original resource, transforming it and packaging it respectively (see van Kranenburg *et al.*, 2010). On the other hand, material and symbolic reproduction of the resource by consumers (tourists) involves, *inter alia*, the transformation of the social structures of destinations by indicating preferences through a choice and interpretation of the offered product. The transformation of symbols of cultural identity into a tourist product without perceived loss of authenticity relies largely on what Berger and Luckmann (1967) call the social 'construction of reality'. Since reality is constructed on stage, the stage imposes its own constraints with the tourists adopting the role of critiques. Therefore, the host has the task of manipulating the tourists' perception of reality in order for it to be in tandem with the constructed reality and fulfil the quest for the experience of authentic Otherness. In Tianxin *et al.*'s (2012) view perceptions of reality are more important than reality itself. This applies to the construction of Otherness.

In order to capture the dynamics of the social construction of Otherness through the presentation of Shona traditional music and dance we focus on the verbal and non-verbal forms of communication between the dancers, dance leader (representing the dance company) on the one hand and the tourist (representing the audience) on the other hand. The analysis of non-verbal communication finds justification in that all kinds of human action, and not just speaking, serve to convey information (see Adegbite, 2010: 133; Akunna, 2008; Flore, 1985: 31, 33).

The Dance Company

The group was founded in September 1977 and Chapungu Cultural Centre was home to the group for almost a decade. The group was composed of 20 young men and women who joined because they had failed to secure employment in the formal sector.

The ages of the group members ranged from 15 to 32 years. All of them had undergone some formal training in dance under Gora, the founder and leader of the group. The training involved watching and imitating the videotaped performances of various dance groups from different parts of Zimbabwe and was conducted at the dance leader's house in Dzivarasekwa, a high-density suburb in Harare. In some cases, certain expert traditional dancers were also invited to come and demonstrate particular dance skills and styles associated with the *muchongowoyo* and *jerusarema* traditional dances.

The Dance Leader

The dance leader's power was based on his in-depth knowledge of Shona culture. Since knowledge is a contested domain, the legitimacy of knowledge-based leadership can only be maintained by resorting to a deliberate mystification of the knowledge base (see Theriou *et al.*, 2009). Gora achieved this through the construction and manipulation of symbols of Shona culture. Symbols became instruments in the construction of images that were targeted at the tourists who were less knowledgeable about the local culture. In that case, leadership was a resource for manipulating the perceptions of the tourist audience in the social construction of Otherness.

Gora was indeed a master of 'impression management' (Goffman, 1959), who straddled the worlds of the host and guest. In his interface with tourists, he determined the aspects of Shona music and dance to present to them. Both the troupe and the audience depended on him for what was to be projected and 'gazed' respectively. From the onset, it is worth noting that Gora was a leader of both the dance troupe and the audience in that besides controlling his group, he was also leading tourists in an exploration of the local culture (see Durand & Calori, 2006). For an insight into the way the leader controlled the tourists' perceptions of the experience we should examine the rhetoric that preceded the actual presentation of the various dances.

Before dances began, usually the dance leader entered the arena to give a brief introduction of the dances that followed. In his opening remarks, he warned tourists: 'The use of audio-recorders and video cameras is prohibited but you are free to take photographs.' The leader did not bother to explain the rationale for the prohibition.

Seemingly, this particular warning disappointed most of the tourists who were anxious to capture the experience live, in order to share it with those at home. The collection of exotic souvenirs is not only a status symbol to tourists but also an integral part of tourist culture and a testimony of the Otherness experienced. The mementos constitute proof of the trip (see Jules-Rosette cited in Harkin, 1995: 652; Kreiner & Zins, 2011; Wilkins, 2009). Appropriation of the tourist experience or object of the gaze through purchase of souvenirs, photographing and video filming is the hallmark of the tourist gaze and an implicit legitimation of the experience of Otherness.

All the dances presented to tourists at Chapungu originated from four ethnic groups collectively referred to as Shona people. These were the Zezuru, Korekore, Manyika and Karanga speaking peoples. Before the dancers appeared on the scene, the leader always made it a point to link the dances about to be witnessed to specific ethnic groups and geographical locations of Zimbabwe. For instance he would say, 'The following dance comes from the Zezuru speaking people of Mrehwa in Mashonaland East Province', '... the Ndau speaking people of Manicaland Province', '... the Karanga speaking people of Bikita in Masvingo Province', from '... the Korekore speaking people of Mashonaland Central Province'.

The above statements are illustrative of the dance leader's rhetorical attempt to convince tourists of the authenticity of the dances they were witnessing and a true expression of Otherness by exploiting the notion of 'territorial identity'. The reference to dialect and geographical location served as 'off-site markers' of authenticity and Otherness. The reference to territorial origin is said to be core to the construction and legitimation of identity (Akunna, 2007, 2008; Andrews, 2011; Buckland, 2001; Kearney, 2013).

Another technique employed by the leader to authenticate and express the Otherness of the various dances was to introduce the dances by their names and functions in Shona society.

The important point to note is, when a dance is presented by its name and function, it assumes a cultural identity and this becomes a defining characteristic of Otherness. In addition, spelling out the social function of each dance conferred a cultural significance to the performance. It made it real and serious as compared to a dance performed for the purpose of mere entertainment. The role of 'function' in authenticating a tourist product is widely acknowledged in literature (see Akunna, 2008: 9; Adegbite, 2010: 133; Littrell et al., 1993: 206).

The concept of Otherness was reinforced by Gora as he presented one ethnic group's dance as having resisted any form of change brought about by the so-called modernity. In a preface to a 'royal' Korekore ethnic group dance the leader said:

> The Korekore people are one of the tribal groups that have consistently resisted change. They have survived the colonial onslaught with all the trappings of modernity it represents.

The statement was intended to engender among tourists an impression of witnessing a dance which was authentic in nature because it had resisted Western values and influence. The Korekore people were cast as a people who shunned change. Thus every tourist audience that came to Chapungu dance arena was told the story of the first white men to come to Zimbabwe. They were told that when the first white man came to Zimbabwe, the ancestors of the Shona people regarded 'the whites as people without knees because they could not see the white man's knees. The trousers made the white man appear as if he had no knees and hence never bent his legs when walking.' Whenever the story was told, it drew a lot of laughter from the audience.

What was peculiar about the story was that it was not linked to any of the dances that followed. Since the story was told to all the tourists who came to the dance arena, it was compelling to explore its significance to the occasion. When asked to explain why he told the story, the dance leader said,

> I just want them to know that we have different pasts. We were contented with our own way of dressing and that's why our fore fathers made fun of the white man's dressing.

Worthy of note was that the story of the first white man was always greeted with laughter from the tourist audience. Perhaps it invoked laughter because principally it was a caricature of the Shona people's ancestors since it portrayed them as a primitive people who regarded a simple pair of trousers as a novelty. In essence, the story of men without knees suggested that the contemporary Shona people were relatively primitive because they lacked a long history of civilisation; hence tourists were expected to perceive the dance as authentic and an expression of Otherness. The suggestion that the leader's speech served to authenticate the presentation by projecting the Shona as people who were close to primitivism seems to be reinforced by the tourists' impressions of the story. When we asked some of the tourists to comment on what the story meant to them, the following various remarks seemed to summarise the opinions of most tourists:

> It simply means your history of civilization lags behind ours (Western). The story goes beyond simple humour. It marks the starting point of civilization in this country. The story gives me a better understanding of the dances I see on stage.

In ethnic tourism, any culture that shuns influence and change is usually perceived as uncontaminated and therefore authentic (Silver, 1993: 308). The dance leader often engaged in rhetorical games to portray Shona culture as generally uninfluenced by Western culture and was therefore on its own a testimony of Otherness. In his talk to tourists he used to say:

> Culturally, Africa is as complete as any other continent. Leave it alone! If ever there was a world festival of cultures, I would say, see them (cultures) as they are, enjoy them as they are, admire them as they are and leave them as they are. Don't disturb them. These differences are beautiful!

When the leader remarked that cultural 'differences are beautiful', he seemed to be reminding tourists that the cultures of hosts and guests were different. In essence, the leader was using rhetorical devices to construct images of 'Otherness' for tourist consumption. The concept of Otherness emphasises the perceived cultural differences between Westerners and non-Westerners. It is the perceived differences that make non-Western cultures appear exotic and therefore authentic.

The next section examines the *Jerusarema* and the *Muchongoyo* dances performed at Chapungu in order to understand how the presence of both the leader and his dancers complemented each other in giving tourists an authentic experience of Otherness of the Shona through dance.

Jerusarema dance

The dance now popularly called *jerusarema* in Zimbabwe is traditionally known as *mbende*. According to tradition, *mbende* was a dance exclusive to married people. Due to the fact that it was generally a sexually suggestive dance, it was only meant for adults, especially married men and women who used it as a public display of sexual skills.

When the first Christian missionaries came to Zimbabwe, they castigated the dance as a 'fornication dance' where only the 'heathens' could participate. According to the dance leader,

> The popularity of the dance necessitated a change of the name by the indigenous people to protect it from being banned by the missionaries. They renamed it *jerusarema* after the Christian city of Jerusalem. The dance later became open to the unmarried and changed its purpose to a courtship dance where boys and girls would form a circle and take turns to dance in pairs of opposite sex.

The males danced to showcase sexual skills, with the dance style reflecting forcefulness, strength and agility. On the other hand, girls' dance mirrored strength and gracefulness as evidenced by dancing with eyes always focused on their hips. In short, it was what Firth (1973: 58) would call a dance of sexual alignment with plainly secular significance.

Out of the several tourists that were asked to comment about the dance, the remarks of one seemed to capture the feelings of many of his colleagues. He said,

> Every society has dances that are suggestive but I found this Shona dance more interesting because of a host of some unique and funny actions that accompanied it.

The important point is, most tourists were convinced of the exotic nature of the dance. Therefore, if exoticism is used as a criterion for judging the authenticity of an experience (Cohen, 1989), then the dancers managed to convince tourists about the authentic nature and Otherness of the dance. This was in spite of the fact that while the real *jerusarema* dance performed in the countryside involved vigorous pelvic thrusts between a man and a woman it was devoid of the 'fainting', fanning and climbing of trees witnessed at Chapungu.

Muchongoyo dance

According to information gathered from a member of Ndau ethnic group elders, the *muchongoyo* dance was originally a ritual dance performed to instil courage among warriors before they left for war. It was usually held the night before the warriors' departure. There are conflicting stories about what really transpired in the dance. Some of the informants recounted that the dance involved a mere simulation of battle situations to induce courage among warriors. However, a majority of them pointed out that in that dance, the warriors' weapons such as spears, axes and others were treated with poisonous herbs. After the administering of the herbs the dance commenced, with women singing and dancing to urge on the warriors' simulations of heroic actions. However, because of the rarity of inter-ethnic wars in contemporary Zimbabwe, nowadays the dance was said to be mainly held to invoke the hunting spirit or honour it for bringing luck to an expert hunter (*muvhimi ane shavi remaisiri*). Commenting on the warrior dances in general, Chiwome in Mutsvairo *et al.* (1996: 118) says 'Although there is little hunting and virtually no inter-ethnic wars today, the dances are still enacted in the veneration of those ancestors who were celebrities in hunting and group defense.'

A look at the Ndau people's warrior dance at Chapungu showed that the elements most popular with tourists were twisting of faces in mock battle, the booming sound of drums, eerie battle cries, a series of somersaults and the heavy pounding of bare feet on the ground. The women's wild gyrations also provided a thrilling experience to the audience. Tourists confessed that they found the heavy pounding of bare feet on the 'hard' ground very exciting and exotic too. It was a novelty to the Westerners. The following remark from a tourist seemed to capture the sentiments of many of his colleagues interviewed about the *muchongoyo* dance:

> I feel a bit nervous. The warrior element looks too real to be just a show to me. However, I keep reminding myself that it's only a show. But the 'warriors' are really dancing themselves to death.

It seems, the seriousness and emotional 'involvement' of the dancers for a while succeeded in making tourists believe they were witnessing a real prewar dance, which in form catered for the tourists' quest for an experience of Otherness.

In sum, survival in cultural tourism requires one to strive to accentuate the differences where they exist and where they do not exist they should be socially constructed. According to Lamont and Fournier (1992) defining the other (or Otherness) requires drawing real or symbolic boundaries. The expression of Otherness at Chapungu Culture Centre was not an exception at all. It exploited both real and symbolic cultural boundaries of the Shona people.

Generally, the construction of images of Otherness by the dance leader was achieved through the manipulation of the social context of the encounter. This was achieved by ensuring that tourists only heard what the leader wanted them to hear and saw only what he wanted them to see (Cohen,

1985: 14). In essence, leadership is a gatekeeper role which entails controlling the interaction process between the dance troupe and the audience to suit the expectations of both parties.

In essence, the dance company presented a 'neo-culture' (Jahn cited in Mudimbe, 1988: 193) or hybrid culture, which was a mixture of local and Western values. The whole idea of acting a people's culture on the stage is a Western phenomenon designed to construct and reconstruct images of the local culture, in accordance with tourist interests. The construction of images of the authentic Otherness to a large extent succeeded because of the involvement of tourists. Tourists became part of the social construction of an authentic Otherness through the positive feedback they gave to the projected images. The feedback was in the form of applauses, laughter, giggles, surprise and other gestures. By so doing, tourists were not only giving meaning to their perceptions but also participating in the construction of meanings and the transformation of images into reality.

Conclusion

The social construction of Otherness primarily depends on the hosts' ability to constantly radiate images of a people whose culture is different from that of the guests. The hosts package their own culture as very dissimilar from that of the guests using a myriad of rhetorical and theatrical skills typifying a situation in which the 'noise' becomes more important than the event. By design, the hosts' rhetoric on cultural differences forms the final and most enduring testimony of the 'Other' more than the visual representations. Of greater significance in the social construction of Otherness is that the hosts require a general knowledge of the culture of the majority of the visitors. This knowledge base is exploited by the hosts to radiate images of cultural differences both rhetorically and visually, with the two operating in mutual reinforcement. Without a good hindsight of the culture of the other, portrayal of images of Otherness by the hosts becomes difficult to achieve. The holistic picture emerging then is that in the social construction of Otherness, the visitors do not necessarily have to taste the 'real' in order to have a satisfactory experience of Otherness. Where the 'authentic Other' is socially constructed, the indicators of the difference between the authentic and inauthentic fade and become indivisible.

References

Adegbite, A. (2010) The impact of African traditional dance. *Journal of Media and Communication Studies* 2 (6), 133–137.

Akunna, G. (2007) *Creativity and the African (Dance) Theatre-Towards a Unified Essence: Perspectives in Nigeria Dance Studies* (pp. 15–25). Ibadan: Caltop Publication (Nigeria) Ltd.

Akunna, G. (2008) Dance as mental therapeutic in the African experience: Beyond the speculation. *African Performance Review London* 2 (1), 9–18.

Andrews, H. (2011) *The British on Holiday: Charter Tourism, Identity and Consumption*. Bristol: Channel View Publications.

Berger, P.L. and Luckmann, T. (1967) *The Social Construction of Reality. Treaties in the Sociology of Knowledge*. London: Penguin Books.

Blatt, B. (1987) *The Conquest of Mental Retardation*. Austin. TX: Pro-Ed.

Buckland, T.J. (2001) Dance, authenticity and cultural memory, the politics of embodiment. *Year Book for Traditional Music* 33, 1–16.

Chiwome, E. (1996) Art and performance. In S. Mutsvairo, N.E. Mberiand, A.M. Masarire and M. Furusa (eds) *Introduction to Shona Culture* (pp. 35–45). Harare: Juta.

Cohen, E. (1985) The tourist guide: Origins, structure and dynamics of a role. *Annals of Tourism Research* 12 (1), 5–29

Cohen, E. (1989) Authenticity and commoditisation. *Tourism Research* (13), 371–386.

Dietvorst, A.G.J. and Ashworth, G.J. (1995) *Tourism and Spatial Transformation*. Oxon: Cab International.

Durand, R. and Calori, R. (2006) Sameness, Otherness? Enriching organizational change theories with philosophical considerations on the same and the Other. *Academy of Management Behaviour* 31 (1), 93–114.

Firth, R. (1973) *Symbols: Public and Private*. London: George Allan and Unwin.

Flores, T. (1985) The anthropology of aesthetics. *Dialectical Anthropology* 10 (1–2), 27–41.

Goffman, E. (1959) *The Presentation of Self in Everyday Life*. Garden City, NY: Anchor Books.

Harkin, M. (1995) Modernist anthropology and tourism of authentic. *Annals of Tourism Research* 22 (4), 781–803.

Kearney, A. (2013) Imaging ethnicities and instruments in Australia and Brazil. In M. Harris, M. Nakata and B. Carlson (eds) *The Politics of Identity: Imaging Indigeniety*. Sydney: UTS Publishing.

Kreiner, C.H. and Zins, Y. (2011) Tourists and souvenirs, changes through time, space and meaning. *Journal of Heritage Tourism* 5 (1), 17–26.

Kroflic, R. (2007) How to domesticate Otherness: Three metaphors of Otherness in the European cultural tradition. *International Journal in Philosophy of Education* 16 (3), 33–43.

Lamont, M. and Fournier, M. (1992) *Cultivating Symbolic and the Making of Inequality*. Chicago: University of Chicago Press.

Levinas, E. (1987) *Time and the Other and Additional Essays*. Pittsburgh: Duquesne University Press.

Lichem, W. (2012) Capacity for Otherness in pluri-identity societies. A presentation at a conference on 'Together Realising the Millennium Development Goals' (pp. 1–8). *A New Vision for Peace and Human Development, UNESCO*.

Littrell, A.L., Anderson, L.F. and Brown, P.J. (1993) What makes a craft souvenir authentic? *Annals of Tourism Research* 20 (1), 197–215.

Mengstie, S. (2011) Construction of 'Otherness' and the role of education: The case of Ethiopia. *Journal of Education Culture and Society* 2, 1–9.

Mudimbe, V.Y. (1988) *The Invention of Africa. Gnosis: Philosophy and the Order of Knowledge*. Bloomington and Indianapolis: Indiana University Press.

Salazar, N.B. (2013) Imagineering otherness: Anthropological legacies in contemporary tourism. *Anthropological Quarterly* 86 (3), 669–696.
Silver, I. (1993) Marketing authenticity in Third World countries. *Annals of Tourism Research* (20), 302–318.
Soyland, S. (2006) The need for Otherness: Spaces of tourism in Nepal. Master's thesis, University of Oslo.
Staszak, J. (2008) Other/otherness. In *International Encyclopedia of Human Geography*. Amsterdam: Elsevier.
Theriou, N.G., Aggelidis, V. and Theriou, G.N. (2009) A theoretical framework contrasting the resource based perspective and knowledge based view. *European Research Studies* XII (3), 177–190.
Tianxin, Z., Takayosh, Y. and Lixin, R. (2012) The performance space for indigenous music in the context of tourism development. A comparative study of Naxi and Hezhe aboriginal areas of China. Soundtracks: Music, Tourism and Travel Abstracts Conference, Liverpool, 6–9 July.
Van Kranenburg, P., Garbers, J., Volk, A., Wiering, F., Grijp, L.P. and Veltkamp, R.C. (2010) Collaboration perspective for folk song research and music information: The indispensable role of computational musicology. *The Journal of Interdisciplinary Music Studies* 4 (1), 17–43.
Wilkins, H. (2009) Souvenirs: What and why we buy? PhD thesis, Griffith University, Australia.

10 Rural Cultural Tourism Development and Agriculture: Evidence from Residents of Mmatshumu Village in the Boteti Region of Botswana

Monkgogi Lenao

Introduction

Regarded as the fastest growing industry worldwide (WTO, 2003), tourism in Botswana has been identified as a potential engine of growth for the national economy (Government of Botswana, 1990; WTTC, 2007). Furthermore, tourism development is seen to be capable of diversifying the country's economy at different levels. At the national level, tourism holds the promise to inject more income into the economy thus reducing the country's dependence on the mineral industry (Government of Botswana, 2013). At the local level, it is promoted for its ability to provide alternative income sources (Mbaiwa, 2008) in addition to the traditional rural industries such as agriculture. To this end, the government has made a deliberate decision to incorporate its development as a strategy in the revised rural development policy (Government of Botswana, 2002). This move has set tourism as a significant livelihood activity in Botswana's rural settings. While wildlife and wilderness have dominated Botswana's tourism offering in the past, culture and heritage products have started entering the tourism market in recent years. This development has introduced tourism to rural areas with little or no wildlife endowment. Invariably, the economic mainstay of these areas has always been agriculture. The chapter explores the relationship between development of rural cultural tourism and agriculture. Using Lekhubu Island as an example, the chapter considers areas of conflict and mutual coexistence between agriculture and rural cultural tourism development.

Rural Cultural Tourism

Rural cultural tourism is an integrative and composite concept. In an attempt to define it, therefore, it is imperative to begin with definitions of its two main constituent concepts, namely cultural tourism and rural tourism. According to Besculides et al. (2002: 303, 304), 'cultural tourism includes visiting historic or archaeological sites, being involved in community festivals, watching traditional dances or ceremonies, or merely shopping for handcrafted art'. Visits targeting features such as natural history, wilderness areas, valued landscapes and buildings have also been identified as constituting cultural tourism (Prentice, 1993). In the same vein, cultural tourists have been identified as those 'interested in the lifestyles of other people, their history and the artefacts and monuments they have made' (Boissevain, 2006: 3). In essence, cultural tourism encompasses travels to places with the view to enjoy and appreciate the entirety of a people's (host community) way of life at the destination.

Rural tourism, on the other hand, is defined as that type of tourism that occurs in rural areas (Molera & Albaradejo, 2007). Borrowing from Lane (1994), they denote the purest form of rural tourism as that 'concerned with tourists who are especially attracted by natural environments and rural culture' (Molera & Albaradejo, 2007: 758). Similarly, Garrod et al. (2006) observe that the success of rural tourism can be measured by its ability to attract visitors to rural areas, satisfy their expectations and ensure that they make return visits to those rural settings. Accordingly, rural tourism may be used as a broad concept capturing a myriad of heterogeneous product mix in rural spaces (Ollenburg, 2006; Tyran, 2007). Accordingly, the attractiveness of natural rural environments, the originality of the rural local cultures as well as the traditional systems of land use and farming, have been singled out as some of the most critical resources for development of rural tourism (Gartner, 2005; Liu, 2006). Therefore, rural cultural tourism denotes a type of tourism that encompasses all the above elements. MacDonald and Jollife (2003: 308) aptly capture this composition as referring to:

> a distinct rural community with its own traditions, heritage, arts, lifestyles, places and values as preserved between generations. Tourists visit

the areas to be informed about the culture and to experience folklore, customs, natural landscapes, and historical landmarks. They might also enjoy other activities in a rural setting such as nature, adventure, sports, festivals, crafts and general sightseeing.

Historically, the rural landscapes have been used exclusively for productive and consumptive industries such as agriculture and forestry. Agricultural economies have been declining in many parts of the world, including Europe (Jaszczak & Zukovskis, 2010), Asia, the US (Gartner, 2005), the Middle East (Fleischer & Pizam, 1997) and sub-Saharan Africa (Kapunda, 2007). In fact, Hazell and Diao (2005: 25) note that, 'Africa has weak institutions for rural development; there is limited irrigation potential and most agriculture must be conducted on depleted soils and under difficult climatic conditions; and world agricultural prices are at historic lows'. This leaves the lives of rural masses in a predicament. As a result, it has become imperative to find other ways and means of augmenting rural economies and helping them sustain multitudes of rural lives dependent on them. Income diversification has been recognised as a viable strategy through which rural households can cope with risks and enhance their survival options (Hazell & Diao, 2005; Turnock, 2002). Rural tourism developments in general, for instance, have been seen to offer rural economies such opportunities as: economic growth, socio-cultural development, and protection and improvement of both the natural and the built environments and infrastructure (Sharpley, 2002). This is the rationale behind community-based tourism in Botswana, in general, and the Lekhubu cultural tourism project, in particular.

Rural Cultural Tourism Development at Lekhubu

Rural tourism has received considerable support owing to its noted potential to help drive development in peripheral areas (Hall & Jenkins, 1998). Giving further impetus to its potential as a suitable tool for diversification of rural economies is the notion that it is founded on the rural cultural landscape, both the natural and human resources found therein as well as the local community's ways of doing things. As Tyran (2007: 122) observes, 'part of the universal appeal of rural tourism rests on the ordinary and everyday happenings of a rural community'. Therefore, it may not be successfully divorced from such traditions as local farming. In fact, it is such traditions that make rural tourism what it is. In other words, this '... conceptual definition embraces notions of local identity, personal contact, closeness to nature, and access to the heritage and residents of the area' (Tyran, 2007: 122–123).

To underscore this point, Turnock (2002: 63) states that 'such tourism can support local infrastructure and conservation programmes, while helping to maintain the viability of small farms and rural communities through additional income, a market for handicrafts and organic farm produce (including traditional foods and drinks) and social interaction for country people and enhance empowerment of women'.

Rural tourism in the southern African region, including Botswana, has taken a community-based approach. Communities have been advised to set up structures (Community Trusts) through which they could establish tourism ventures in order to derive benefits from the natural, cultural and heritage resources found within their localities. In Botswana, this has taken the form of community-based natural resources management (CBNRM) projects where communities practice sustainable utilisation of these resources while also sharing in their management. According to Phuthego and Chanda (2004: 60), 'CBNRM is designed to alleviate rural poverty by empowering communities to manage resources for long-term social, economic and ecological benefits'. Zuze (2009) notes that at the end of 2009 about 85 such projects were operational countrywide. While the majority of these are wildlife-based, other projects are also managed, such as veldt products, culture and heritage resources. The non-wildlife-based CBNRM projects in Botswana are relatively newer compared to their wildlife-based counterparts, mainly since CBNRM in Botswana was first conceived for wildlife management (Phuthego & Chanda, 2004; Thakadu, 2005). One example of rural cultural tourism development is found at Lekhubu Island in the Makgadikgadi pans landscape. This tourism is owned and run by residents of Mmatshumu village located 45 km to the south of the Island and on the fringes of the Makgadikgadi pans.

Geographic description of the study area

Lekhubu Island itself is a rock outcrop jutting out of the flat Makgadikgadi pans. This feature covers an area of approximately 60 ha, and is characterised by a relatively small colony of large baobab (*Adansonia digitata*) and African star chestnut (*Sterculia africana*) trees. Besides a variety of birds, snake and rodent species that may be spotted at Lekhubu, there is no presence of larger animal species at this island. Culturally, Lekhubu Island has always been important to the residents of Mmatshumu village and adjacent settlements owing to its rich culture and heritage endowment. Different communities that have lived in and around Mmatshumu have used the area for ancestral worship, traditional and cultural rituals as well as for hunting purposes (Campbell, 1991).

At a national level, the significance of Lekhubu Island stems from archaeological remains dating back centuries. Among the archaeological features found at the island is a popular shrine, comprising walls of piled up stones measuring between 70 and 100 metres in length. The shrine is believed to have served as an initiation centre for boys during the 17th century (Campbell, 1991). In addition, there are some heaps of stones arranged in a grave-like formation. Legend has it that the heaps mark the graves of those boys who died from haemorrhage during circumcision. An alternative explanation is that at the end of the initiation, boys built those heaps as a sign of their presence. Furthermore, some pottery fragment remains and beads made from shells and bones have also been found at the site. In view of the archaeological, cultural and heritage significance, the government of Botswana declared Lekhubu Island a National Monument in 1938.

In 1997, the community of Mmatshumu village set up the Gaing'O Community Trust (GCT), through which they established and now run a cultural ecotourism project at Lekhubu Island. This was a reaction to the increasing concern among community members that Lekhubu island resources were being exploited by non-residents for economic benefits at the exclusion of Mmatshumu residents. The concern was given further impetus by the recognition that unregulated tourism activity at the island would ultimately destroy its historical, cultural and natural resources. There were already indications that some individuals and groups of people were, in fact, removing some artifacts from the site without permission – a practice in contravention of the code of conduct applicable to visitors to sites of this nature. Since its inception, the GCT has managed to build and furnish an office in the village, acquire a four-wheel-drive vehicle, computers, cameras, tents, and build a reception block and ablutions at the site as well as create some camping sites. The GCT also employs about eight members of the community on a permanent basis.

Methods of Data Collection

This chapter is derived from a broader qualitative study conducted among the Mmatshumu village community in the Boteti sub-district of Botswana. This wider study investigates the conditions and circumstances of Lekhubu Island as a community-based tourism site and implications for future development. For this chapter, secondary data from different sources including official reports, published journal articles, books, booklets and newsletters were used. Primary data used in this chapter were collected from

Mmatshumu village, Lekhubu and other adjacent settlements between January and June 2012. The researcher conducted in-depth interviews, key informant interviews and focus group discussions with different members of the community.

A total of five key informant interviews targeting those members of the community with first-hand experience in running the Lekhubu project were conducted. These included current and former Board chairpersons, secretary, board member, manager and field assistant(s). In addition, five in-depth face-to-face interviews targeting tribal and civic leaders were conducted. Among those interviewed were the current and retired village chiefs, the Village Development Committee (VDC) chairperson and secretary as well and the area councillor. The rationale for conducting these key informant and in-depth interviews was to solicit the views of those information-rich subjects of the study (Kitchin & Tate, 2000). As expected, these informants were able to shed light on some intricate issues relating to the day-to-day running of the Lekhubu project, including challenges and opportunities as well as the vision for the future.

On the other hand, ordinary members of the community were targeted through focus group discussions (FGDs). A total of seven such FGDs were conducted. Each of the FGDs comprised between eight and 12 members except one that was made up of 16 participants. FGD compositions were made according to gender and/or age. Group descriptions included: mixed gender-mixed age, mixed gender-youth, male only-youth, female only-youth, male only-adults and female only-adults. Invitation to participate in the FGDs was extended to any member of the community aged 16 years or above. Members were visited at their own homesteads and invited to a specific location within the village. Some were met while walking around the village, while others were found at the local *shebeens* (taverns), clinics, *Ipelegeng* (a Botswana government programme aimed at reducing poverty through engaging people in manual labour or supplementary works such as grass and tree clearing in the village) sites and any other areas in the village where people could be found gathered during the day. The scheduled times and type of gender or age group expected were explained to the potential participants beforehand.

The data gathered were analysed thematically (Hoggart *et al.*, 2002). This process began with allocating each interview and FGD transcript a unique colour code for ease of identification and differentiation. Then, an analysis template was created following the themes on both the interview and FGD guides. Information was then copied from the transcripts and pasted to the template in accordance with the guides. Leininger (1985) supports this as it allows for bringing together various fragments of information to enable better comprehension. Following Aronson's (1994) advice, the emerging themes were pieced together to form a comprehensive picture of the collective views. The following sub-section presents the findings and discussions of issues raised during both the key and in-depth interviews as well as the FGDs.

Areas of Competition and Conflict

Presence of cattle at Lekhubu Island

It is critical to note that the nearest cattle posts (*meraka*) to Lekhubu Island are located between 5 and 8 km away. This is significant given the fact that livestock from these cattle posts is free range and Lekhubu Island is not fenced. Inevitably, therefore, these have unfettered access to the island any time of the day or night. This has been identified as an area of concern by both the tour guides based at the site as well as some tourists whose views were recorded in the message book kept at the island. While tourists' messages generally appreciated the near pristine beauty and serenity of the place, there were numerous references to the presence of cattle being something of a nuisance at the island. On the one hand, some entries made reference to the actual presence of cattle at the island, as exemplified by the following appraisal: 'Great spot, maybe too many cattle around here.' On the other hand, even when cattle were not physically present at the time, the mere signs of them having been there upset some tourists. One of the tourists quipped, 'sorry about cattle trails plus loads of dung around this place'. When discussing these sentiments with the guides resident at Lekhubu Island, they expressed their concern with the increasing numbers of cattle coming to the island both during the day and at night.

According to the guides, the cattle disturb visitors by exhibiting curious behaviour around the tents while the occupants are relaxing. Some cattle wear bells around their necks and these can cause tremendous amounts of noise, thus disturbing the natural serenity of the island. The cattle also disturb the stones meant to demarcate campsites and fireplaces, thus making the camping grounds look untidy. They also drop dung which smells, attracts flies and also leaves the area unpleasant. According to the guides at the site, the responsibility of keeping the cattle away from the camping grounds rests with them, and they feel this takes a lot of their time and energy which they would have otherwise reserved for giving service to the tourists. As one guide expressed, they already feel more like herdboys (*badisa*) as opposed to guides. Notwithstanding the complaints, it is also interesting to note that the island presents one of the few areas within the vast Makgadikgadi salt pans with grazing opportunities for cattle. Since there is no barrier or fence between the nearest cattle posts and the island, the place essentially makes up part of the communal grazing area for Makgadikgadi livestock.

Competition for labour force

It has been noted elsewhere that development of tourism at any destination presents income-generating opportunities to different members of the community. Most importantly, tourism, compared to other sectors of the economy, has the potential to ensure quicker and higher returns to those involved. This short-term promise for income provides an allure for people traditionally involved with other sectors of the economy to abandon such activities and move towards tourism. This situation is particularly apparent at destinations where the primary economic mainstay would have been agriculture, such as in the Mmatshumu area of the Makgadikgadi pans landscape. It is a stated objective of the GCT to encourage local community members to come up with tourism-related activities, products and services that can be sold to the tourists who come to visit Lekhubu as a means of empowering them, as well as diversifying both the island's tourism offering and the livelihood options of the residents.

Among some of the activities, products and services envisaged to be provided by the local communities are traditional song and dance (presenting traditional dances of the different ethnicities residing in Mamatshumu village, namely Basarwa, Bakalaka and Bahurutshe), pottery, different types of crafts and local cuisine. From the interviews and FGDs, there appears to be a common consensus that facilitating for community members to provide entertainment and produce crafts for sale as rural cultural tourism at Lekhubu would be a good initiative. In fact, the community and local leadership generally believe that this would provide a good opportunity for the locals to be involved in the business of tourism and enjoy the benefits thereof. On the basis of this, therefore (and while the magnitude may not be determined), it could be possible that development of rural cultural tourism at Lekhubu Island has the potential to compete for labour with such areas as agriculture. For instance, opportunities to participate in cultural tourism activities (e.g. craft making and traditional song and dance) at Lekhubu Island could provide alternative sources of livelihood for persons currently making a living from working in both arable and pastoral agriculture. Opportunities to earn income through tourism may be more attractive than agriculture if residents consider income from tourism-related activities to be any higher or easier to obtain.

Potential Areas of Synergy and Complementarities

Market for agricultural products, goods and services

Among others, the development of tourism in rural areas has been seen to complement agriculture by creating or increasing markets for locally produced agricultural goods. In a study carried out in the village of Bigodi, in Uganda, it was observed that tourism improved the market for agricultural produce in two ways. According to Lepp (2007), tourists who visit Bigodi consume products from local farmers. Furthermore, some locals who have found employment in the bourgeoning tourism industry have been left with little time for agriculture and as a result have opted out of the industry. These locals who do not produce for themselves have resorted to buying from those who are still farming. In each case, the market for locally produced agricultural goods has improved (Lepp, 2007). Similarly, Mellor and Lele (1973) also found that demand for locally produced goods and services increased significantly as a result of linkages developing between farm and non-farm activities during India's Green Revolution. Haggblade et al. (2005: 167) observe that: '... nonfarm income – perhaps from mining or rural administrative centers – generates demand for local agricultural products. Hence, the commonly observed truck farming that grows up around rural and urban towns.'

While discussing this topic, Mmatshumu residents expressed optimism that their traditional foods would appeal to the tourists coming to visit Lekhubu. One of the residents identified *seswaa* (pounded meat) and *phane* worm (a caterpillar commonly found on the *colophospermum mopane* tree from which its name is derived) as very popular delicacies that would naturally be of interest to visitors. According to her,

> whenever, *phane* or *seswaa* are being sold alongside other types of relish in restaurants in Letlhakane, Orapa or any of the major towns and cities, these two would run out first. I think white people are fascinated by these more than anything else. That is why I think development of our tourism project would bring us market for our traditional food. (Female 33, FGD participant)

During another FGD, comprising elderly women, traditional dishes were talked about extensively. They observed that the highest numbers of tourists visiting Lekhubu are normally witnessed during the dry winter time, which coincides with the time when the women would have just finished processing their harvests from the fields, thus giving them a great opportunity to supply the needed ingredients for the popular dishes of the area. In essence,

Mmatshumu residents place a strong emphasis on the potential linkages between agriculture and tourism development in their area.

General infrastructural development

Hazell and Diao (2005) recognise that public investment in rural infrastructure is a prerequisite for growth in both agriculture and other non-farm activities. In other words, diversified rural livelihoods options would be facilitated by development of rural infrastructure. This point may also be used to argue that diversified rural livelihood options would also necessitate investment in rural infrastructural development, thus promoting improved rural economies. Development of rural infrastructure is necessary for 'strengthening rural-urban demand linkages' (Hazell & Diao, 2005: 24) for both agriculture and non-farm economic activities including tourism. A conglomeration of agricultural and non-farm economic activities would give rural landscapes the necessary bargaining power for increased public spending.

In Mmatshumu village, youth who participated in the FGDs expressed hope that with the advent of rural tourism development within their area they will have access to more infrastructure such as telephone and cellular communication coverage as well as the internet. They cited the presence of a tele-centre in their village as a sign of better things to come in this area. It was noted that, with such infrastructure available, it would be possible for local farmers to market their products to the potential consumers. In fact, reference was made to the fact that the same Lekhubu Island project website could be used in part to promote farm products from the Mmatshumu area. They also recognised that the construction of a 13 km stretch of tarred road connecting Mmatshumu village to the Orapa/Letlhakane/Francistown junction, generally attributable by locals to the development of tourism in their area, would provide local farmers with easier access to markets in major villages and towns.

Tours to working cattle posts

Flyman (2003) identifies a possible point of convergence and complementarity between tourism and pastoral agriculture in Botswana's Okavango Delta. He argues that, more than just meat and milk, cattle also form a critical part of the rural landscape. This argument is further extended to include the tourists visiting the Okavango to enjoy scenery enhanced by the presence of cattle without necessarily paying for that added value. Flyman (2003), therefore, proposes that agricultural holdings such as working cattle posts may be officially incorporated into the tourist visits to the area to enable tourists to enjoy and learn about the country's farming methods while possibly immersing themselves in the activities of the farm operation.

Safari companies can contribute in bridging the gap by developing complementary products with local communities to make destinations more attractive to tourists, extend the length of stay and provide employment and other income benefits to the poor whose way of life constitutes an important part of the holiday experience (Flyman, 2003: 19). While discussing this issue, some members of the community mentioned the possibility of encouraging tourists to visit working cattle posts in the area, although most of them did not know how this could be achieved. One of the FGD participants suggested that a similar approach to that involving heritage trails (connecting heritage sites) could be adopted. According to him, there are a number of cattle posts in the vicinity of Lekhubu and tourists may be interested in learning about life at these cattle posts while also engaging in some of the activities, such as milking the cows and so on. He argued that deliberate decisions should be made to create routes that link up these cattle posts and Lekhubu Island.

Conclusions

This chapter sought to investigate the relationship between agriculture and rural cultural tourism development in Lekhubu Island. Using the views from residents of Mmatshumu village, the chapter provides evidence that there is a potential for both positive and negative coexistence between agriculture and tourism development at Lekhubu Island. Among the possible or existing areas of conflict between tourism development and agriculture is the fact that tourists and care takers of the Island decry disturbances caused by cattle from adjacent cattle posts. The presence of cattle at the site is considered something of a nuisance. This is an interesting challenge given that the island constitutes a part of the wider communal grazing land for local cattle posts. Another potential area of conflict was inferred on the basis of the residents' enthusiasm about rural cultural tourism development in Lekhubu. This potential area of conflict may involve competition over labour between agriculture and tourism development.

In terms of the potential synergies (positive relationships), development of tourism presents potential opportunities for local communities to produce and sell part of their farm produce to the bourgeoning tourism sector, where it would be used to prepare traditional dishes. While producers would benefit from selling their produce, the value of the area's traditional food may also be enhanced as cultural tourism items. Residents of Mmatshumu also believe that development of tourism in their area presents an opportunity for local infrastructural development which would in turn provide opportunities for farmers to easily access outside markets for their own farm products. In addition, residents have identified the potential for working cattle posts to act as tourism attractions in their own right. To this end, a suggestion was made that tourists could be encouraged to visit the cattle posts and partake in the daily activities therein, such as milking the cows.

References

Aronson, J. (1994) A pragmatic view of qualitative analysis. *The Qualitative Report* 2 (1), 1–3.
Besculides, A., Lee, M.E. and McCormick, P.J. (2002) Residents' perceptions of the cultural benefits of tourism. *Annals of Tourism Research* 29 (2), 303–319.
Boissevain, J. (2006) Coping with mass cultural tourism: Structure and strategies. *Gazeto Internacia de Antropologio* 1 (1), 2–11.
Campbell, A.C. (1991) The riddle of the stone walls. *Botswana Notes and Records* 23, 243–249.
Fleischer, A. and Pizam, A. (1997) Rural tourism in Israel. *Tourism Management* 18 (6), 367–372.
Flyman, M.V. (2003) Bridging the gap between livestock keeping and tourism in the Ngamiland District, Botswana. Concept paper prepared for ACCORD, Centre for Tourism and Natural Resources Management, Gaborone, Botswana, December.
Garrod, B., Wornell, R. and Youell, R. (2006) Re-conceptualizing rural resources and country-side capital: The case of rural tourism. *Journal of Rural Studies* 22, 117–128.
Gartner, W.C. (2005) A perspective on rural tourism development. *The Journal of Regional Analysis and Policy* 35 (1), 33–42.
Government of Botswana (1990) *Tourism Policy Paper No. 2 of 1990*. Gaborone: Government Printers.
Government of Botswana (2002) *Revised National Policy for Rural Development*. Gaborone: Government Printers.
Government of Botswana (2013) *Mid-term Review of NDP 10: NDP 10 towards 2016*. Gaborone: Ministry of Finance and Development Planning.
Haggblade, S., Hazell, P. and Reardon, T. (2005) The rural non-farm economy: Pathway out of poverty or pathway in? In *The Future of Small Farms: Proceedings of a Research Workshop*. Wye, UK, 26–29 June.
Hall, C.M. and Jenkins, J. (1998) The policy dimensions of rural tourism and recreation. In R. Butler, C.M. Hall and J. Jenkins (eds) *Tourism and Recreation in Rural Areas* (pp. 199–203). Chichester: John Wiley.
Hazell, P. and Diao, X. (2005) The role of agriculture and small farms in economic development. In *The Future of Small Farms: Proceedings of a Research Workshop*. Wye, UK, 26–29 June.
Hoggart, K., Lees, L. and Davies, A. (2002) *Researching Human Geography*. London: Arnold.
Jaszczak, A. and Zukovskis, J. (2010) Tourism business in development of European rural areas. Management theory and studies for rural business and infrastructure development. Research papers. Nr. 20(1).

Kapunda, S.M. (2007) Beyond the impasse of African industrial development: The case of Botswana, Tanzania and Zambia. *African Development* (XXXII) 4, 99–108.
Kitchin, R. and Tate, N.J. (2000) *Conducting Research in Human Geography: Theory, Methodology and Practice*. Harlow: Pearson.
Lane, B. (1994) What is rural tourism? *Journal of Sustainable Tourism* 2 (1), 7–21.
Leininger, M.M. (1985) Ethnography and ethno-nursing: Models and modes of qualitative data analysis. In M.M. Leininger (ed.) *Qualitative Research in Nursing* (pp. 33–72). Orlando, FL: Grune and Stratton.
Lepp, A. (2007) Residents' attitudes towards tourism in Bigodi Village, Uganda. *Tourism Management* 28, 876–885.
Liu, A. (2006) Tourism in rural areas: Kedah, Malaysia. *Tourism Management* 27, 878–889.
MacDonald, R. and Jolliffe, L. (2003) Cultural rural tourism evidence from Canada. *Annals of Tourism Research* 30 (2), 307–322.
Mbaiwa, J.E. (2008) Tourism development, rural livelihoods and conservation in the Okavango Delta, Botswana. PhD thesis, Texas M&A University.
Molera, L. and Albaladejo, I.P. (2006) Profiling segments of tourists in rural areas of south-eastern Spain. *Tourism Management* 28, 757–767.
Mellor, J. and Lele, U.J. (1973) Growth linkages of the new food grain technologies. *Indian Journal of Agricultural Economics* 18 (1), 35–55.
Ollenburg, C. (2006) Family tourism in Australia: A farm family business and rural studies perspective. PhD thesis, Griffith University.
Phuthego, T.C. and Chanda, R. (2004) Traditional ecological knowledge and community-based natural resources management: Lessons from a Botswana wildlife management area. *Applied Geography* 24, 57–76.
Prentice, M. (1993) *Tourism and Heritage Attractions*. London: Routledge.
Sharpley, R. (2002) Rural tourism and the challenge of tourism diversification: The case of Cyprus. *Tourism Management* 23, 233–244.
Thakadu, O.T. (2005) Success factors in community based natural resources management projects mobilization in Northern Botswana: Lessons learnt from practice. *Natural Resources Forum* 29 (3), 199–212.
Turnock, D. (2002) Prospects for sustainable rural cultural tourism in Maramures, Romania. *Tourism Geographies* 4 (1), 62–94.
Tyran, E. (2007) Regional and traditional products as an important part of rural tourism offer. *Oeconomia* 6 (3), 121–128.
World Tourism Organization (WTO) (2003) *Tourism Market Trends-Africa*. Madrid: World Tourism Organization.
World Travel and Tourism Council (WTTC) (2007) *Botswana: The Impact of Travel and Tourism on Jobs and the Economy*. London: WTTC.
Zuze, C.S. (2009) *Community Based Natural Resources Management in Botswana: Practitioners' Manual*. Gaborone: Department of Wildlife and National Parks.

11 From Hunting-Gathering to Hospitality? Livelihoods and Tourism Use of Bushman Paintings in the Brandberg Mountain, Namibia

Renaud Lapeyre

Introduction

Lying within the Tsiseb area, in the Erongo region, north-west Namibia, the Brandberg Mountain, 2573 m high at Koenigstein peak, is home to 900 sites containing over 50,000 rock paintings (Figure 11.1). Since the country's independence in 1990, its cultural value has attracted over 10,000 visitors per year (Lapeyre, 2009). Most of these come and visit the world-renowned White Lady paintings by trekking three hours both ways in the mountain, whereas others undertake a three-day trek in order to reach the peak while visiting bushman paintings on the way. As a result of its archaeological and tourism importance, the Brandberg Mountain has been listed as a World Heritage Site since 2002.

Local communities, from hunter-gatherers to pastoral nomads, have long inhabited the Brandberg Mountain area (Kinahan, 1991). For the last decades the Daures daman and #Aodaman clans of the Damara tribe have depended on the area for water and grazing resources; over this time, importantly, the mountain has maintained cultural and heritage value among local people with persistent traditional and spiritual beliefs.

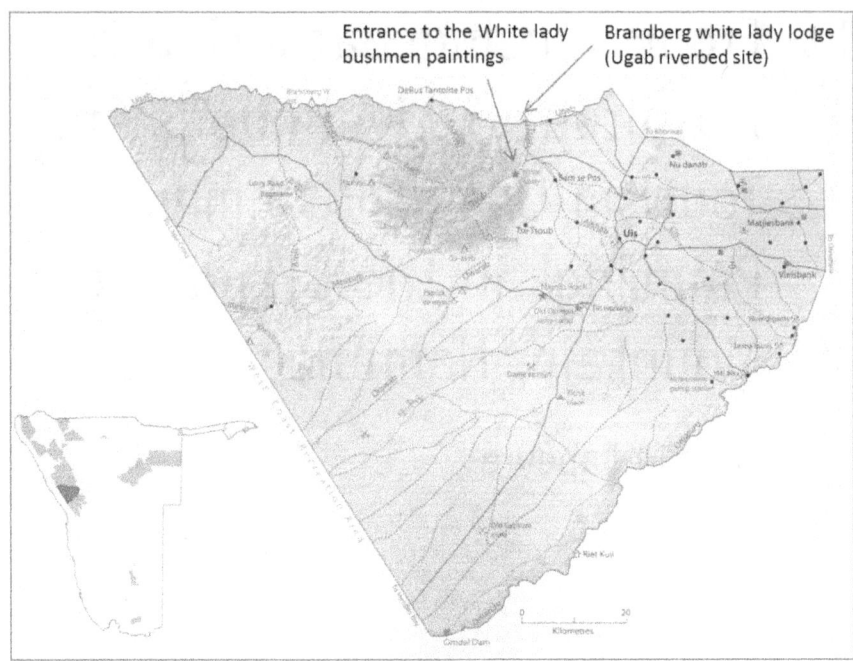

Figure 11.1 The Tsiseb conservancy and the Brandberg Mountain
Source: Author, adapted from Namibian Association of Community Based Natural Resource Management Support Organisations (NACSO).

For this reason, although since August 1951 the mountain area has been declared a National Monument under state supervision by the National Monument Council, local traditional authorities have for some time informally endorsed young farmers to preserve the mountain, including archaeological sites, and look after visitors coming to visit the White Lady painting site (Lapeyre, 2009). Since 1993, cultural tourism activities on the mountain have thus been carried out and organised by a local community project where approximately 15–20 local farmers have become tour guides, guiding tourists daily to the White Lady paintings. The project infrastructure was at first very limited. From 2000, however, several donor and non-governmental organisations (NGO), including the European Union Namibia Tourism Development Programme (EU-NTDP) and the Namibian community-based tourism association (NACOBTA), have funded infrastructure improvement on-site (protecting the paintings, reception, interpretation centre, etc.) and significantly invested in technical assistance as well as institutional capacity building.

At the larger regional scale, in 2001 the local community also formed and registered the Tsiseb (*Hada Huigu*) conservancy, stretching over 800,000 ha. The conservancy includes the former mining town of Uis, part of the Messum crater, the Ugab river and wetland area, and the Brandberg Mountain (see Figure 11.1). Institutionally, the conservancy is a territorial unit over which a legally recognised group of local people is devolved partial use rights over wildlife and tourism activities. To be registered, the group must define clear and legitimate boundaries, set a list of members, agree on a constitution, elect a representative committee and chairperson, and finally design a natural resource management plan (Jones, 1998, 2003). Once declared a conservancy, the latter is then allowed to control its natural resources and decide how to use and benefit from these. In particular, it is able to sell commercial and trophy hunting rights, as well as develop commercial tourism activities on its territory, whether carrying these out itself or partnering with private operators.

Building on this new legal framework, the Tsiseb conservancy has since its inception benefited from wildlife and tourism resources. Though bushman paintings on the Brandberg Mountain remain the main attraction in the conservancy, the presence of desert-adapted elephants roaming in the Ugab river and other endemic flora (*Weltwischia*) nicely complements the tourism product which attracts visitors to the area.

Box 11.1 Cultural village tourism in Namibia: The case of Helvi Mpingana Kondombolo Cultural Village

Mary-Ellen Kimaro and Hilma Joolokeni Nengola

Cultural heritage sites aim to reflect the values and identities a nation or a group of people would like to pass on to its future generations. Namibia's policy on arts and culture (MYSC, 2001: 3) states that the goal of arts and culture is to promote unity in diversity and give all Namibians a sense of identity and pride. The national policy (MET, 2008: 10) on tourism reiterates that: 'Government will promote and encourage the experience of local culture, traditions and customs and ensure that culture is not inappropriately exploited.'

Cultural villages can be seen as a specific form of cultural heritage and tourist attraction found in both rural and urban areas. They are currently perhaps one of the main forms of cultural heritage tourism in Namibia. According to Silvester (2013), four types of cultural village in

(*continued*)

Box 11.1 Cultural village tourism in Namibia: The case of Helvi Mpingana Kondombolo Cultural Village (*continued*)

Namibia have emerged. One is based on the belief that tourists want 'authenticity' and therefore tour operators organise township tours in urban areas and visits to 'real' local villages in rural areas. The second model is based on the idea of 'staged authenticity' where tourists come into contact with performers who present a 'front stage' of their village and culture. In contrast to that they have a 'back stage' where they actually live and carry out their day-to-day activities. The living museum of the Ju/'Hoansi San in Tsumkwe is an example of staged authenticity in Namibian cultural villages. The third model of cultural villages is based on a more interactive approach with an aim to encourage cultural dialogue; tourists attend workshops to learn about traditional skills and learn how to use local plants (e.g. Munyondo gwaKapande cultural village in the Kavango Region). The fourth model of Silvester's (2013) typology aims to display the diverse traditional homesteads of the different ethnic groups of Namibia.

The Helvi Ya Mpingana Kondombolo Cultural Village (Plate 11.1) represents the fourth type of cultural village in Namibia. It was established in the town of Tsumeb in 1995. The emphasis of the village is on the different architectural styles and building materials traditionally used by different Namibian ethnic groups. In addition to homesteads, the cultural village has a gallery containing a permanent display of crafts and a curio shop offering Namibian craft products. The Helvi Ya Mpingana Cultural Village is a good example of the nature and challenges that heritage sites face in tourism in Namibia. These challenges are briefly discussed against the critical success factors of a cultural village based on literature (see Carlsen *et al.*, 2008; Gyimothy & Johns, 2001; Ho & McKercher, 2004; Hughes & Carlsen, 2010; Mattsson & Praesto, 2005) and manager and staff interviews and a visitor survey (N = 35) conducted at the site in 2013. The critical factors include: (a) Set and agreed objectives and concept; (b) Effective human resource management; (c) Quality and authenticity of products and experiences; (d) Effective marketing and market research; (e) Planning for product differentiation, life cycles and value adding; (f) Financial planning; (g) Visitor flows and location; (h) Engaging expertise in conservation and promotion; and (i) Interpretation.

Plate 11.1 The entrance to the Helvi Ya Mpingana Kondombolo Cultural Village
Source: Ellen Kimaro (1999).

(a) *Set and agreed objectives and concept*: Interviews indicated that the initial idea for a cultural village in Tsumeb was created by a group of people that visited Norway in the 1990s, where the concept of open-air museums was experienced. With a desire to see a similar set-up in Namibia displaying the diverse traditional 'tribal' homesteads, a committee made up of various stakeholders, including the community, was set up. The committee agreed on the concept and initiated the construction. The main objective for the initial development phase of the cultural village was to preserve the values and identities of the Namibian nation for future generations and to have a platform that would serve as an outdoor classroom, where domestic and international tourists could visit and learn.

(b) *Effective human resource management*: For the successful operation of the cultural village a combination of skills in preservation, conservation and business management would be an added advantage to ensure success (Hughes & Carlsen, 2010). The cultural village has only five full-time staff: two women as senior supervisors and three men who act as tour guides and maintenance staff. Enriching the

(*continued*)

Box 11.1 Cultural village tourism in Namibia: The case of Helvi Mpingana Kondombolo Cultural Village (*continued*)

staff component with additional employees with skills in preservation, conservation and business management would benefit the village. The use of volunteers with these skills could also be explored.

(c) *Quality and authenticity of products and experiences*: The quality of the experience at the cultural village is reliant on the ability to nurture the interest of tourists and challenge them intellectually. It is also related to the tourists' expectations on the authenticity of the place. According to the surveyed visitors, almost one-half (49%) of them felt that the cultural village was authentic, whereas one-third (30%) implied that they did not know.

(d) *Effective marketing and market research*: Internet marketing of the cultural village and an interactive website would probably increase visitation numbers. Visible signage that is easily identifiable would also be an added advantage. The initial exploratory study done by the authors was the first undertaken since the establishment of the cultural village.

(e) *Planning for product differentiation, life cycles and value adding*: The cultural village needs to tap into its uniqueness in terms of being able to display the diverse ethnic homesteads and tell stories related to the different cultures located in one area. Value adding can be achieved by including new products in the future, and planning for tourism features in the local economic development (LED) strategy.

(f) *Financial planning*: Initial funding for construction was received from the Norwegian twin town of Elverum, but after the construction phase the existing funding could not sustain the village. In order to avoid its closure, the municipality took over the operational costs, management and maintenance in 1996. The municipality of Tsumeb carries out the financial planning under a tight budget. It foresees handing over the cultural village's operations to the community in the future. Currently the cultural village charges visitors a small nominal fee that does not cover the cost of maintaining the village.

(g) *Visitor flows and location*: The cultural village is found along one of Namibia's tourism routes from Etosha National Park. However, it lacks clustering or proximity to other nearby attractions. The village has not been able to establish links with tour operators using this route.

(h) *Engaging expertise in conservation and promotion*: There is a need to strike a balance between running the cultural village as a commercial venture and conserving and preserving its heritage values.
(i) *Interpretation*: Expertise in providing effective interpretation leads to a positive tourist experience. A majority of surveyed visitors (63%) were satisfied with the interpretation and agreed that the guides at the village had knowledge about what they were presenting. However, this still leaves room for improvements and training of staff.

To conclude, the Helvi Ya Mpingana Cultural Village has touristic potential if the mentioned critical success factors are revisited and reconsidered in its future operations. The village is important in strengthening the country's cultural and heritage tourism as part of the Namibian tourism product. The objective of preserving cultural heritage values and identities for future generations and balancing this with running the cultural village on sound business principles could create a win–win situation.

References

Carlsen, J., Hughes, M., Frost, W., Pocock, C. and Peel, V. (2008) *Success Factors in Cultural Heritage Tourism Enterprise Management (Report to Sustainable Tourism Cooperative Research Centre)*. Gold Coast, QLD: Sustainable Tourism CRC. See www.crctourism.com.critical success factors for heritage tourism (accessed 22 July 2014).

Gyimothy, S. and Johns, N. (2001) Developing the role of quality. In S. Drummond and I. Yeoman (eds) *Quality Issues in Heritage Visitor Attractions* (pp. 243–266). Oxford: Butterworth-Heinemann.

Ho, P. and McKercher, B. (2004) Managing heritage resources as tourism products. *Asia Pacific Journal of Tourism Research* 9 (3), 255–266.

Hughes, M. and Carlsen, J. (2010) The business of cultural heritage tourism: Critical success factors. *Journal of Heritage Tourism* 5 (1), 17–32.

Mattsson, J. and Praesto, A. (2005) The creation of a Swedish heritage destination: An insider's view of entrepreneurial marketing. *Scandinavian Journal of Hospitality and Tourism* 5 (2), 152–166.

Ministry of Environment and Tourism (MET) (2008) *A National Tourism Policy for Namibia*. Windhoek: Republic of Namibia.

Ministry of Youth, Sports and Culture (MYSC) (2001) *Namibia's Policy on Arts and Culture*. Windhoek: Republic of Namibia.

Silvester, J. (2013) Trading in tradition: The development of cultural villages in Namibia. See www.maltwood.uvic.ca/cam/publications/conference_publications/Jeremy%20Silvester.pdf (accessed 23 June 2014).

While the community project's guiding activities in the Brandberg Mountain have been studied elsewhere (see Lapeyre, 2009, 2010), this chapter aims at analysing tourism accommodation activities closely dependent on cultural tourism in the Brandberg Mountain. Specifically, the intent is to assess impacts of such cultural tourism-based accommodation on local inhabitants' livelihoods and level of empowerment. To do so, this chapter will build on both an institutional analysis (Ostrom, 1990) as well as the sustainable livelihood approach (Scoones, 1998). The next section will present institutional arrangements that have up to now governed accommodation activities within the Brandberg Mountain area, and associated conflicts that have arisen. After that livelihood impacts will be assessed from such activities with results from a household survey conducted with employees working at the nearest tourism lodge (the Brandberg White Lady Lodge, see Figure 11.1). Finally the results will be discussed.

Accommodating Cultural Tourists through Partnering with the Private Sector: The Brandberg White Lady Lodge and Associated Conflicts

Situated approximately 3 km from the entrance of the mountain where guided tours to the bushman paintings start, the Ugab riverbed site has long been used for camping and accommodation purposes by cultural tourists coming from, or going to, the White Lady painting site in the Brandberg Mountain. Although erratically, water is available on-site, while essential shade is provided by Ana trees and vegetation.

From the Ugab wilderness community campsite (UWCC) to the joint-venture lodge: Institutional design of the partnership lodge

Although the Ugab riverbed site was initially used privately by a white entrepreneur (since 1997), in 1999 the local community (the emerging Tsiseb conservancy) was willing to benefit from this campsite, legally situated in a communal zone. Progressively, Namibian NGOs, including the Rural Institute for Social Empowerment (RISE) and NACOBTA, supported by outside donors such as the World Wildlife Fund for Nature (WWF) and the Unites States Agency for International Development (USAID), came to assist the community project through institutional and capacity building as well as business advice. In 2001, NACOBTA chose to contract a small private tourism operator to manage the community campsite, now called UWCC, on behalf of the conservancy. However, conflicts arose along racial misunderstandings while rights, responsibilities and duties were not divided clearly enough between the company and the Tsiseb conservancy. The company eventually decided to terminate their management contract with the conservancy. As a result, NACOBTA decided to hire a local young man (also a tour guide at the Brandberg Mountain) to manage operations at the UWCC; yet local tourism and business expertise was still lacking and employment conflicts broke out. In 2002, an external manager was again hired to coordinate the project, but the institutional set-up was once more too weak. Results were disappointing and the community got frustrated. At this point a new institutional arrangement had to be found again.

Because of such recurrent organisational and business issues in UWCC – also observed in several other similar community tourism projects in Namibia – support agencies (NACOBTA, RISE, etc.) and their donors (USAID, WWF, etc.) changed the paradigm regarding tourism management in rural areas. Acknowledging their limited organisational and financial capacity and their lack of skills and expertise in tourism, communities were incited to lease out their rights over prime tourism sites to skilled and experienced private entrepreneurs and companies interested in investing in accommodation in community areas (conservancies). The Ugab riverbed site was here no exception and as a result the Tsiseb conservancy decided to call for private-sector involvement in the development of a lodge there.

In order to find potential investors to develop accommodation close to the Brandberg Mountain, NACOBTA, through funding by WWF and USAID, launched a national process where the Ugab riverbed site was tendered (together with three other sites situated each in a different conservancy). After the site was visited in September 2002, tenders for rights over the tourism site were submitted in November 2002. Formally, tenders had to include a financial as well as a technical offer. Whereas the financial offer proposed a lease fee (a percentage of net turnover) annually paid to the conservancy, including a guaranteed minimum fixed lease fee, the technical offer included a business plan, an environmental proposal and finally an empowerment proposal. The latter stated a minimum commitment to training and career advancement and to contracting local suppliers and service providers. Further, it committed to hire local community members in operational middle and senior management positions after a certain period of operation.

Eventually, only one company actually tendered to take over the operation of UWCC and build a lodge in the Ugab riverbed site. The interested investor was a professional builder who had a long-term interest in the area.

Although the adjudication team assessing the bid documents did score both the environmental and empowerment proposals badly and therefore recommended declining such a bid and allowing for more time to find alternative better investors, the Tsiseb conservancy nevertheless awarded the tender to this investor in November 2002. As a result, negotiations started within a joint-venture committee which comprised the investor, five members from the conservancy as well as three representatives from NGOs supposedly playing the role of honest brokers.

The contract was signed in March 2003 in the form of a lease agreement. In the latter, the conservancy, holding rights over the Ugab riverbed site, lets the approximately 10 hectare lodge site and area to the lessee, the private investor, for the purpose of building and operating a tourism lodge, called the Brandberg White Lady Lodge, and associated tourism activities. It further grants the lessee an Exclusive Development Zone (EDZ) to develop tourism infrastructure, and rights to traverse the entire conservancy area for tourism purposes (e.g. game drives). The lease agreement was set to continue for a renewable 10-year-period, and was actually renewed in 2013. On contract termination, the investor shall hand over the infrastructure (immoveable assets) to the conservancy free of charge. Besides, according to the contract agreement, the investor must strictly implement and adhere to its proposed empowerment plan: local residents should be hired as a priority and provided with training sessions, and 'at least 80% of operational middle and senior management at the lodge will be from the local community by the 3rd year'.

In order to monitor compliance with such respective parties' rights and duties and discuss potential contractual issues, the contract sets up a joint management committee (JMC), consisting of three representatives of each party which meets not less than once every quarter. Importantly, this committee is responsible for proposing and nominating candidates for employment and discussing any labour issue arising at the lodge.

Institutional instability: Conflicts over the partnership

Although a win–win solution in theory, such a partnership contract, developing accommodation to promote cultural tourism in the Brandberg Mountain, has actually proved conflicting on the ground. First, problems rapidly arose regarding the investor's exclusivity to use the tourism site and the area. On the one hand, although the contract prohibits such movement, local farmers' cattle, goats and donkeys have up to now freely come into the tourism lodge, disturbed tourists and broken water points. On the other hand, outside overland tours and groups often illegally camp around the Brandberg Mountain even if it is in the investor's EDZ, without paying any fee. In this case, as stated by the investor, 'everybody loses: [the investor] lose[s], the conservancy loses, it's a lose-lose situation' (Interview of the lodge owner and manager, July 2007).

Second, the issue of empowerment was rapidly raised by the community. According to the latter, the investor has not fulfilled its commitments for formal training and career advancement, while no real management position was proposed to local staff (see the next section for further evidence). Third, labour conditions have from the beginning been problematic (again see the next section for further evidence). Conservancy representatives at the JMC mentioned several times the issue of employment procedures within the partnership (hiring and firing), as well as the lack of signing of formal employment contracts and of registration of workers with social security. Labour practices have similarly been very tense at the lodge. Unfair, violent and sometimes racist treatment of employees was reported both by employees themselves and tourists staying at the lodge; besides, staff quarters were said by the conservancy to look like a 'blikkies dorp' (a 'squatter camp' in Afrikaans). Further, though stipulated in the contract, such labour issues are not discussed at the JMC level and left to the investor's discretion. In total, even if the number of problems has reduced recently, the partnership remains uneasy and unstable, jeopardising the product quality (e.g. manifested by the number of bad reviews of the lodge made by visitors on the internet) and in the end the cultural tourism experience.

Community Impacts of Cultural Tourism in Tsiseb: From Limited Empowerment to Livelihood Opportunities

In order to disentangle impacts of cultural tourism in the Brandberg Mountain at the household level, we further administered household questionnaires to local community members in the Tsiseb area. While socio-economic impacts of cultural tourism for tour guides (guiding tourists to rock paintings) have been analysed elsewhere (Lapeyre, 2010), we focus here on such impacts for households involved in accommodation activities in the Brandberg White Lady Lodge.

To this end, we interviewed 30 lodge employees (18 male and 12 female) either in September 2006 (21 surveys) or July 2007 (9 surveys). Overall, the average age of staff was around 30 years old, and the average level of education was approximately grade 9. Only two employees completed a diploma after grade 12. Geographically, most of the staff came from the urban settlement of Uis (53%), whereas 37% of respondents were living in rural settlements very close to the lodge site. Overall, 80% of all employees interviewed reported farming (at least part-time) within the conservancy. Three types of impact from cultural tourism activities in the area are of specific interest to us: employment, empowerment and livelihoods.

Employment impacts: From pastoral nomads to low-skilled jobs?

Positions filled in the Brandberg White Lady Lodge were mostly low-skilled, with low responsibility. In 2006, positions included 5 waiters or receptionists, 8 cleaners (lodge and campsite, now operated by the private company), 1 gardener, 1 chef, 3 general kitchen staff (lodge), 2 laundry ladies, 3 maintenance persons and finally 1 guide (only casually). Wages paid to employees (excluding food rations), ranged from N$700 (1 Euro = N$10 at the time) for lower positions (12 respondents, including 6 cleaners, 2 kitchen ladies, 2 laundry ladies and 2 maintenance guys), N$1600 for the chef, and N$1700 for one senior barman and receptionist. The average monthly wage was estimated to be approximately N$855 (excluding food rations).

When compared to average monthly income per capita in rural areas, estimated at N$410 (NPC, 2006), and the salary of farm workers (less than N$3 per working hour, see Lapeyre, 2010), this income proves quite significant. However, job security and stability seemed to be problematic at the lodge. As per September 2006, out of 23 employees, only 5, all senior, had signed a formal contract (22%). Similarly, few respondents could be sure whether they were registered with social security (6 workers could certify they were registered, 5 stated the contrary and 13 could not answer). In this context, employees have been working under high contractual uncertainty and job instability. Consequently, staff turnover has been quite high. In July 2007, out of 24 employees working at the project in September 2006, 10 had already left after less than a year (42%) while only 14 remained employed. Hired and fired at the investor's discretion, employees rarely work for long at the lodge. In September 2006, the average length of stay for 17 workers employed at the lodge was 14 months (out of 30 possible months, as the lodge opened in June 2004). Strikingly, almost half of these 17 workers (eight) were very new employees (less than seven months tenure).

Working, as well as living, conditions at the lodge might explain such staff turnover and instability. First, working conditions were reported to be difficult. Indeed, while 57% of employees declared they were quite happy with working conditions (see Table 11.1), some respondents actually reported they were afraid of criticising for fear of losing their job. For instance, a

Table 11.1 Working conditions as reported by employees interviewed

Working conditions	Number of respondents	% (N = 30)
Managers have bad manners (shouting and colonial type)	7	23
No precise job description	3	10
Salary issues	3	10
Insecure: 'pack and go' style	3	10
Work is hard and stressful	2	7
Good	17	57
Does not answer	2	7

Source: Author.

41-year-old laundry woman, mother of four children, told us: 'there are problems but we don't want to lose our bread. That's why we are quiet.' On the contrary, almost one-quarter of respondents mentioned bad behaviour from the lodge investor to explain bad working conditions, as illustrated by a 21-year-old male staff member: 'they [managers] come and shout. They still have that colonial type of rules. We don't benefit from the fruits of independence here.' Further, 10% cited low salaries and non-payment of overtime and working Sundays as reasons for discontent; 10% complained that their job description was very unclear and that managers used them for many unrelated activities; and finally, 10% felt insecure as managers were very quick to fire them, even if they previously made no mistakes. Staff members have to strictly obey if they do not want to lose their main source of income. In the words of a waiter and receptionist, 'the owner talks to the people as "if you don't want to work, you must leave". It's a "do or die" type of thing.'

Living conditions were further cited as a main problem for those who stayed at the staff quarters. Although 20% of employees declared they were happy with living conditions at the lodge, significantly one-half complained about the lack of electricity at the staff quarters or the limited comfort provided (e.g. lack of privacy in the small rooms to be shared, no hot water, only open-air showers for the whole staff).

Empowerment effects: Partnership for cultural tourism or business as usual?

At the employee level, empowerment at the lodge is very limited. First, very few employees from our sample have obtained a promotion or a change in their job position and description, and few salaries were increased. Since they started at the lodge, one-quarter of the respondents had been promoted while one-quarter had received an actual wage increase.

Although clearly stipulated in the contract, capacity building was not undertaken by the investor: no formal training sessions have been provided to the staff. On the contrary, the investor showed employees how to do their tasks in a non-systematic and very informal way. Neither certificates nor official recognition have been obtained by employees. As a result, employees are not empowered and left with few skills and limited further opportunities to get a job in another company. A kitchen woman who started in 2004 illustrated well the point when she said: 'here is no empowerment; there is not anything that (…) you learn. Here there is no certificate, or something else, to look for another job.'

Similarly, responsibilities and decision-making powers are seldom shared with employees, who continue carrying out low-skilled tasks. Receptionists, though theoretically at middle-management position, are not empowered to take any decisions or be involved in the management of the lodge. For instance, a young qualified woman at reception (with a diploma) told us: 'I don't have more power. I am sitting here. I am here. I am hoping tomorrow is a better day. What I am thinking, [is] you have to start with a lower level and get to a better level [but] is there really empowerment [here]? (…) I know computer but she [the manager] is the whole day with the computer. I could be with the computer in management position, but I am not…Is it empowerment?'

Decision-making powers, responsibilities and skilled tasks are concentrated in the investor's hands. Far from benefiting from the partnership, employees only obey the investor and have very low power, as in any other privately owned and managed lodge. Asked about potential empowerment, a 22-year-old general worker answered: 'I don't feel it. Power lies in the hand of managers (…). [We] benefit from salary but we don't have any kind of much decision-making power (…).'

Unfortunately, community empowerment from the partnership is also not felt at the broader conservancy level. Indeed, as already mentioned, the conservancy is very seldom involved in the process of hiring staff. When interviewed, only 30% of respondents declared that their hiring was a joint process between the conservancy and the investor within the JMC. The conservancy is similarly very seldom involved *via* the JMC in firing and disciplinary procedures. According to employees, the conservancy is not really supportive of solving staff problems (registration with social security, signing of contracts, staff quarters, and disciplinary questions). On the contrary, a 35-year-old employee who was fired for a drinking problem explained: 'in the time I was fired, I complained to the conservancy, but they (the chairman) just told me to go back to [the manager/investor]. There was no meeting to solve the problem of the firing. They (the JMC) did not come together.'

In total, community empowerment and involvement in monitoring the partnership seem very limited. Rather than building a strong partnership, cultural tourism in Tsiseb seems to be run under a weak lease agreement. In the very illustrative words of a receptionist and waiter at the lodge:

> The involvement of the conservancy? It just gets money from the business. (...) Their involvement is not that clear. (...) We don't see them that much. (...) There is no empowerment. If they would have empowered us, thus, we would have learnt, and would have had training. But the owner [the investor] takes decisions. We don't take decisions about the lodge. It does not seem the conservancy has a role here. (...) It's kind of a private lodge, run by private people. It does not seem it is run by the conservancy or local people.

Sustaining rural households' livelihoods through income generation

While the previous section has provided evidence of limited career advancement, capacity building and empowerment generated by cultural tourism in the Tsiseb area, we also investigated the potentially beneficial effects of wage income distribution on households' livelihoods. Wages paid by the Brandberg White Lady Lodge first represent a very critical source of cash income in a rural area where very few economic alternatives exist (the tin mine closed in 1990) and extreme aridity only allows for subsistence farming.

As shown in Table 11.2, by analysing employees' previous livelihood paths and trajectories, the socio-economic changes brought by the lodge prove very significant. On the one hand, working at the lodge allowed two respondents (young males) who had just finished their studies (Grade 12 or polytechnic institute) to get their first job and gain experience. On the other hand, before being employed at the lodge, 57% of respondents were unemployed: 37% were just staying at their family farm looking after goats, 13% were staying at home in an urban settlement (e.g. Uis) and 7% were operating a very small business on their farm (e.g. selling eggs and chickens). In this context, their new jobs at the lodge allowed these households to buy food, pay for children's school fees and uniforms, and to cope with unforeseen external shocks, such as drought, injuries and funerals.

Furthermore, cash income distributed through this tourism activity is most of the time non-substitutable and irreplaceable. Half of the respondents

Table 11.2 Employees' previous professional occupations

Previous occupation	Number of respondents	% (N = 30)
Unemployed – staying at farm	11	37
Unemployed	4	13
Unemployed – staying at farm/small business at farm	2	7
Hospitality sector	3	10
UWCC (community campsite)	2	7
School and diploma	2	7
Mining	1	3
Other	4	13
n/a	1	3

Source: Author.

fully rely on tourism for their households' livelihoods (Table 11.3): 30% only receive their wage at the lodge and 20% generate cash from their wage together with tips they get when they sing (choir) for tourists at the lodge. Further, although one-third of respondents report running small businesses to complement their livelihoods, most of these actually generate very limited alternative income by selling tobacco, cigarettes, sweets, chips and beers at the staff quarters or renting a pool table. In all cases, salaries from the lodge remain the major and almost exclusive source of income for employees. In this context, this makes a huge difference, as illustrated by a 41-year-old woman: 'while I was at ! Gaus [the closest farm to the lodge], there was no way to make money. Now that I am here, I get some income at the end of the month and I can help myself.'

Now endowed with a source of cash income, employees are able to support their households. When interviewed, respondents reported they

Table 11.3 Lodge revenue as the main source of income for employees

Other source of income	Number of respondents	% (N = 30)
None – only wage from the lodge	9	30
Singing for tourists	6	20
Having a small business	9	30
Owning a pool table in a bar	1	3
n/a	5	17

Source: Author.

supported, in cash or in-kind, an average of 3.21 dependents (excluding themselves). Consequently, we estimate that in July 2007 (total staff = 30), the lodge financially generated benefits for around 126 people in the area (employees and their dependents).

Practically, most employees (54%) supported their dependents by simply handing out cash when they were in need (e.g. hospital visits or funerals) (Table 11.4). This money remains limited and not regularly paid, yet it helps dependents to cope with unforeseen shocks and related expenses. One-third of employees further reported buying food for their family and dependents, 11% paid for clothes and necessary items (e.g. blankets), while, finally, almost one-quarter declared they paid for education for their children and their relatives' children. In the latter case this provides young people who are depending on lodge employees' wages with the opportunity to build their human capital for future development (future labour force); this is, we contend, probably one of the most significant and sustainable socio-economic impacts from this tourism partnership.

Employees' own spending patterns (once they have helped their dependents and relatives) also provide evidence of a positive socio-economic change in rural livelihoods brought about by the tourism partnership (Table 11.5). While 78% of respondents declared they could buy necessary personal items for themselves, one-third reported they could now use their regular wage (paid each month) in order to buy bigger equipment on credit (furniture, such as beds; electronic devices, such as CD players) and then repay their loans.

More significantly, wages received from their tourism employment allowed people to build and maintain their financial, physical as well as human capital. Three respondents (11%) declared they tried to save (financial capital); two declared (7%) they could buy donkeys or goats with their salary (physical capital); and 7% could pay a herder to look after livestock

Table 11.4 Employees' support for their relatives and dependents

Type of support	Number of respondents	% (N = 28)
Give money	15	54
Pay for food	9	32
Pay school/hostel/kindergarten fees/books/uniforms	6	21
Buy toiletries/clothes/shoes/blankets	3	11
Not specified	7	25

Source: Author.

Table 11.5 Employees' personal spending

Expenses	Number of respondents	% (N = 27)
Buy personal things	21	78
Repay loans	9	33
Pay for a house rent	4	15
(Try to) save	3	11
Pay a herder	2	7
Buy livestock/donkeys	2	7
Repay studies	1	4

Source: Author.

and cattle (investment) at their farm (while they are away working at the lodge); finally one (4%) could repay the amount he owed for his four years of study (human capital).

Participation in Cultural Tourism versus Rural Poverty Alleviation: A Discussion

As it is based on local spiritual assets and traditional knowledge, informally held de facto by autochthonous inhabitants, cultural tourism in southern Africa is often alleged to better benefit local communities by actively involving and empowering them. This chapter has investigated such a hypothesis in the case of a lodge accommodating tourists visiting bushman paintings in Brandberg Mountain, Namibia.

Although designed as a win–win community/private-sector partnership, the Brandberg White Lady Lodge actually proves problematic. The conservancy is barely involved in lodge management decisions and employees feel disempowered and insecure. Far from displaying characteristics of co-management, the institutional design resembles a weak lease agreement where the local community remains a passive landowner, only renting its land and natural resources, with no opportunity for further capacity building, investment in human capital and empowerment. In this context, we contend, the potential for socio-economic sustainability at the local level, whereby the conservancy would eventually take over the efficient operation of the lodge after the contract ends in 2023, seems elusive.

Such results showing the currently limited possibility for community empowerment and participation in the tourism sector are not isolated. In the case of the Tsiseb conservancy, Newsham (2007) also showed that because knowledge actually lies with government, NGOs and specific private-sector actors, traditional authorities as well as community members are not genuinely participating. Thus, community members remain de facto recipients, dependent on external assistance for capacity building. In two other Namibian contexts, Hoole (2010) finds similar trends: while the tourism partnership in the Torra conservancy has undoubtedly proved quite successful in training community members, power-sharing remains unbalanced between local inhabitants and the private operator; also, in the Ehi-rovipuka conservancy, local people's empowerment and participation is clearly absent.

Against this backdrop, institutional redesign of such so-called partnerships should be a priority in order to really empower communities. In other places in Namibia, such as #Khoadi-//Hôas, financially assisting the conservancy so that the latter invests and actually owns lodge infrastructure has created leverage for significantly genuine involvement and empowerment (Lapeyre, 2011). Yet, going beyond empowerment and participation issues, such accommodation activity in Tsiseb has clearly provided local households with the opportunity to improve their livelihoods in a semi-arid context where few other income-generating economic activities exist (Brown, 2004). As demonstrated in many other studies in Namibia (Snyman, 2012a; Jones et al., 2013), Botswana and Malawi (Snyman, 2012b), our results strongly indicate that wages distributed at the lodge allow employees to repay their study loans, invest in livestock and pay for their children's school fees. Hence, building their households' financial and physical capital through employment in tourism will eventually help local farmers to better respond to unforeseen shocks and improve their socio-economic resilience. Acknowledging such positive socio-economic outcomes for rural poverty alleviation, one might contend that seeking to further improve communities' participation in tourism activities may in fact significantly increase transaction costs and, thus, undermine income distribution objectives. In the end, one might have to choose between efficiency and procedural equity, facing trade-off rather than synergy.

References

Brown, C. (2004) Namibia's conservation paradigm. Use to conserve versus protect to conserve. *Conservation* 2004/5, 3–5.

Hoole, A.F. (2010) Place-power-prognosis: Community-based conservation, partnership and ecotourism enterprise in Namibia. *International Journal of the Commons* 4 (1), 78–99.

Jones, B.T.B. (1998) Namibia's approach to CBNRM: Towards sustainable development in communal areas. Scandinavian Seminar College, African Perspectives of Policies and Practices Supporting Sustainable Development in Sub-Saharan Africa, Windhoek, September.

Jones, B.T.B. (2003) Policy, institutions and practice. The impact of Namibian policy and legislation on rural livelihoods (*WILD Project Working Papers* 25). Windhoek: DFID and MET.

Jones, B.T.B., Davis, A., Diez, L. and Diggle, R.W. (2013) Community-based natural resource management (CBNRM) and reducing poverty in Namibia. In D. Roe, J. Elliott, C. Sandbrook and M. Walpole (eds) *Biodiversity Conservation and Poverty Alleviation: Exploring the Evidence for a Link* (pp. 191–205). London: John Wiley & Sons.

Kinahan, J. (1991) *Pastoral Nomads of the Central Namib Desert: The People History Forgot*. Windhoek: Namibia Archaeological Trust, New Namibia Books.

Lapeyre, R. (2009) Rural communities, the state and the market: A new-institutional analysis of tourism governance and impacts in Namibian communal lands. PhD thesis, University of St Quentin-en-Yvelines, France.

Lapeyre, R. (2010) Community-based enterprises as a sustainable solution to maximise tourism impacts at the local level? The case of indigenous projects in the Tsiseb Conservancy. *Development Southern Africa* 27, 757–772.

Lapeyre, R. (2011) The Grootberg lodge partnership in Namibia: Towards poverty alleviation and empowerment for long-term sustainability? *Current Issues in Tourism* 14 (3), 221–234.

Newsham, A. (2007) Knowing and deciding: Participation in conservation and development initiatives in Namibia and Argentina. PhD thesis, Centre of African Studies, University of Edinburgh.

NPC (2006) *National Household Income and Expenditure Survey 2003/2004. (A preliminary Report)*. Windhoek: National Planning Commission, Central Bureau of Statistics.

Ostrom, E. (1990) *Governing the Commons. The Evolution of Institutions for Collective Action*. Cambridge: Cambridge University Press.

Scoones, I. (1998) Sustainable rural livelihoods: A framework for analysis (*IDS Working Papers, 72*). Brighton: Institute of Development Studies.

Snyman, S.L. (2012a) Ecotourism joint ventures between the private sector and communities: An updated analysis of the Torra Conservancy and Damaraland Camp partnership, Namibia. *Tourism Management Perspectives* 4, 127–135.

Snyman, S.L. (2012b) The role of tourism employment in poverty reduction and community perceptions of conservation and tourism in southern Africa. *Journal of Sustainable Tourism* 20 (3), 395–416.

12 Emergence of Cultural Tourism in Southern Africa: Case Studies of Two Communities in Botswana

Masego Monare, Naomi Moswete, Jeremy Perkins and Jarkko Saarinen

Introduction

Cultural tourism is a growing social, cultural and economic activity (UNWTO, 2012; WTTC, 2007, 2013) with a relatively long history (see MacDonald & Jolliffe, 2003). In recent decades, cultural tourism has increasingly expanded to new geographical areas, such as southern Africa and its remote parts, where tourism per se may still represent a relatively new sector of economy (see Mbaiwa & Sakuze, 2009; Saarinen, 2010, 2011). Cultural tourism can be regarded as a 'cultural experience' and according to Richards (1997: 24) it can be defined as the movement of persons to cultural attractions away from their normal place of residence, with the intention to gather new information and an experience to satisfy their cultural needs.

In general, cultural tourism is not limited to material manifestations such as monuments and objects that have been preserved over time, but includes socio-cultural, economic, natural heritage and environmental aspects and traditions of societies. Thus, cultural tourism therefore consists of intangible cultural elements which include language, tales, myths and history, music, songs, dances, traditional motifs, literature, rituals, customs and lifestyles, for example (Keitumetse, 2005; UNWTO, 2008), that are current or not, that have the capacity to inform the present about the past

(Jones, 2009; Mbaiwa, 2005, 2011). Further, cultural tourism encompasses living expressions and the traditions that countless groups and communities worldwide have inherited from their ancestors and transmit to their descendants, in many cases orally, especially in African contexts.

In southern Africa, cultural tourists arrive to the region to seek an 'authentic' and 'deep' experience with local people and cultures (see McKercher & du Cros, 2002; Saarinen *et al.*, 2014). Due to the supposed connections with a search for authenticity and meaning, it is usually seen as a

Box 12.1 Authenticity in tourism

Jarkko Saarinen

The concept of authenticity is widely used in tourism studies. In general, it refers to something that seems genuine or original to us. In cultural tourism it can relate to tangible (e.g. crafts and other physical cultural elements) or intangible (e.g. festivals, traditions and performances) cultural resources and related experiences. In tourism studies the conceptual basis originates from the early discussions by Boorstin (1964) and MacCannell (1973). Boorstin criticised mass tourism for creating pseudo-events and commodifying destination cultures, thus eroding the originality and authenticity of host cultures. MacCannell (1976) further developed the conceptual thinking in his seminal work *The Tourist: A New Theory of the Leisure Class*. He approached tourism from within a framework of modernity and its evolution in Western societies, by saying that the progress of modernity depends on its very sense of non-authenticity, i.e. alienation from authenticity. In order to cope with alienation in everyday life, MacCannell considered that people seek authenticity and related experiences as tourists.

According to MacCannell (1976), tourists searching for authenticity are also actively creating and recreating the structures of authenticity in tourism which modernity has caused to vanish elsewhere. Thus, authenticity does not refer only to a 'nostalgic past' but it has a geographical dimension: in tourism it can still be found in faraway places. Indeed, in tourism advertising, for example, the past, authenticity and real experiences are often located at the peripheries of the modern world, where indigenous or other ethnic groups (seen as untouched by modernity) are 'situated'. However, MacCannell divided authenticity into two different levels or contexts: 'front stages' and 'back stages' in destinations.

According to him, tourists usually experience the touristic front stage that is opened up and constructed for them to visit (e.g. cultural villages), while the back stage is where local people actually live in their contemporary culture.

Although it has had a relatively long history in tourism studies, the idea of authenticity is still a highly contested and debated issue in research. Critical voices are asking if there is actually anything authentic in tourism; i.e. if all elements of cultural tourism, for example, are more or less constructed for visitors to experience and consume (see Cohen, 1988; Selwyn, 1996). From this perspective, the authenticity of indigenous people's 'genuine' handicrafts, for example, can be questioned, as traditionally indigenous people have not made those artifacts for sale or exchange (at least not for tourists!); they are made mainly for their own use. Thus, many indigenous or ethnic crafts sold in southern African craft markets (Plate 12.1) represent indigenous or ethnic cultures as much as tourism culture.

There are different conceptualisations of authenticity and its role in contemporary tourism. Wang (1999: 365) has stated that 'even if toured objects are totally inauthentic, seeking otherwise is still possible, because tourists can quest for an alternative, namely, existential authenticity to be activated by tourist experience'. Based on this line of thinking, Wang has made a fruitful categorisation of the different notions of authenticity, namely objective, constructive and existential authenticity. Objective authenticity refers to 'the authenticity of original' (Wang 1999: 352) and a museum kind of originality (a 'true realness' of artifacts and events). In contrast, constructed authenticity refers to the authenticity projected onto toured objects and events by tourism producers or tourists. As a socially constructed process, the nature of authenticity is also negotiable. Thus, it can be emphasised 'through cultural representations of reality' in the processes of tourism, as initially noted by MacCannell (1976: 92). Existential authenticity relates closely to the tourists' experiences. It may have nothing to do with the (objective) authenticity of toured objects. Therefore, Wang (1999) emphasises that it is important to make a distinction between the authenticity of experiences and the authenticity of toured objects, which would better allow for the utilisation of authenticity in tourism by involving the dimensions of existential authenticity rather than the authenticity of objects in cultural tourism products. However, this relativism and flexibility that is empowered by the existential notion of authenticity should not lead to touristic

(continued)

Box 12.1 Authenticity in tourism (*continued*)

Plate 12.1 Craft market development at Hartbeespoort Dam, South Africa
Source: Jarkko Saarinen.

products that represent the destination cultures in inappropriate and unethical ways from the perspective of the host culture.

References

Boorstin, D. (1964) *The Image: A Guide to Pseudo-Events in America*. New York: Atheneum.
Cohen, E. (1988) Authenticity and commoditisation in tourism. *Annals of Tourism Research* 15 (3), 371–386.
MacCannell, D. (1973) Staged authenticity: Arrangements of social space in tourist settings. *American Journal of Sociology* 79 (4), 589–603.
MacCannell, D. (1976) *The Tourist. A New Theory of the Leisure Class*. New York: Schoken Books.
Selwyn, T. (ed.) (1996) *The Tourist Image: Myths and Myth Making in Tourism*. Chichester: John Wiley & Sons.
Wang, N. (1999) Rethinking authenticity in tourism experience. *Annals of Tourism Research* 26 (2), 349–370.

unique form of tourism (Moswete & Dube, 2013). It often differs from the so-called mass tourism, because cultural tourists are seen to attempt to go beyond idle leisure and return enriched with knowledge of other places and other people, although this may sometimes involve gazing at or collecting the commodified essences of others (see Saarinen, 2009). In the same vein, Richards (2003) has echoed that unlike beach or mountain tourism, for example, which have to have certain elements of physical geography, cultural tourism can be developed almost everywhere because every place has its unique culture which may be difficult to copy elsewhere.

The purpose of this chapter is to explore the emergence of cultural tourism in Botswana with specific reference to cultural village tourism at two case sites: Bahurutshe and Motse cultural villages. The chapter attempts to answer the following research questions: (a) How and why have the cultural villages been created?; (b) What is the ownership of the villages?; (c) Do both villages offer cultural tourism in the context of Botswana or as understood in the country?; (d) What type of activities and services are offered?; and (e) Who are the key stakeholders?

Cultural Tourism

Brief overview

Cultural tourism can represent an ideal vehicle for community-based and rural tourism development where people experience direct economic benefits from visitors in their villages (Jones, 2009). In this context, it has been occasionally celebrated as a potential saviour of many communities around the world because of its capacity to generate hard currency, new income and jobs (Cole, 2007; McKercher & du Cros, 2002; Stephen, 1992). However, the negative elements of cultural tourism can emerge when it becomes increasingly commercialised and a non-locally driven development activity (see Gordon, 1990; Mbaiwa, 2011; Weiman, 1993).

Indeed, several studies have revealed benefits and costs of cultural tourism (Mbaiwa, 2011; McKercher & du Cros, 2002; Moswete & Lacey, 2014). Cultural tourism encourages awareness of cultural history of individual ethnic groupings (MacDonald & Jolliffe, 2003; Stephen, 1992). It serves as a kind of 'reminder' that is used to lead observers into the cultural past to see the present from another viewpoint (Ivanovic, 2008; Smith, 2009). The tourist, through cultural tourism, reverses the accelerating experiences of leisure time and seeks a contemplative journey of adventure into the past (Bachleitner & Zins, 1999) or contemporary times. By experiencing culture, visitors and

Table 12.1 Examples of cultural heritage attractions in southern Botswana

Name of attraction	Location	District/region
Toutswemogala Iron Age Site	Near Palapye	Central
Matsieng's Foot Prints	Rasesa	Kgatleng
Goo-Moremi Gorge	Tswapong Hills	Central
Domboshaba Ruins	Near Masunga	North East
Logaga la ga Kobokwe	Molepolole	Kweneng
Thamaga pottery	Thamaga	Kweneng
Lentswe la Baratani	Otse	South East
Old Palapye Church Ruins	Malaka	Central

Source: BTDP (2000); Ford et al. (2010); GoB (2001).

tourists are entertained and also learn about the places visited, communities, heritage and the cultural landscape. Many economies have embraced culture and history in their tourism sectors because of the 'new tourist', whose interest or motivation is to experience many cultures – tangible and intangible, of different nations, in the first and developing world (see Table 12.1).

Cultural tourism in Botswana

Cultural tourism is a relatively new development in Botswana (Saarinen et al., 2014). Hunting safaris have occurred for over a century, with photographic tourism in areas with high densities of wildlife (i.e. northern Botswana) and/or remote wilderness areas (Makgadikgadi Pans/Kalahari) characterising tourism in Botswana over the last 30–40 years. Ferrar (1995: 32) has pointed out, in relation to an IUCN (International Union for Conservation of Nature) visitor survey, that:

> The typical visitor: around 40 years old, English speaking, a professional, earning about US$40,000 per annum. Half come from the United Kingdom or North America and a quarter from South Africa. They travel with family or friends averaging four per group and spend around US$2,700 on the holiday, US$1,000 of which is spent in Botswana. Their opinions are that the wildlife, scenery and lack of people are great but in the Parks, the facilities, staff service, roads and lack of interpretative material could be improved. Their very clear advice is to 'keep it wild', resist civilising and developing the Park, make visitor facilities adequate and functional but keep them basic. Scenic wilderness, peace and quiet, and learning something to improve their understanding is what they expect from their visit.

The interpretative and informative parts of visitors' experiences in Botswana are elements where cultural tourism can play a leading role. Visitors to the Makgadikgadi Pans have in fact long been drawn by the diversity of Stone Age artefacts that can be found there and its deep connectivity with the Kalahari San who lived there, rather than the populations of large herbivores which on the Pans themselves are in fact few and far between. Cultural tourism that is independent, or far from the key wilderness and wildlife areas in Botswana does of course offer a new challenge in terms of drawing visitors away from the well-beaten 'Big Five Game Viewing' circuits that continue to dominate tourism in Botswana. At the same time, it presents the opportunity to bring tourism to hitherto neglected parts and cultures of Botswana and diversify not only the tourism product itself but also those who are able to benefit from the industry.

Cultural Villages in the Context of Botswana

Policy background

Cultural villages are commonly known as and/or associated with tourism (Monare, 2013). In other countries, such as South Africa (e.g. Shakaland Zulu cultural village) and Namibia (e.g. Damara cultural village), they are sometimes known and operate as 'living museums'. In Botswana, they are also referred to as 'cultural lodges' as they offer touristic services needed by visitors (e.g. lodging and local cuisine). As leisure and tourism operations they often tend to be frequented by international and domestic 'white' tourists (individuals and group tours) more than any other type (Monare, 2013). Generally, a cultural village is a special place which shows the culture of the community in which it has been built (see Ford *et al.*, 2010: 78; Mbaiwa, 2011). Cultural villages are specific attractions as they symbolise the way of living of local people, hence visitors learn about the culture of the people (see Gurung, 1995; Saarinen, 2009). A cultural village can manifest itself in a number of ways (materially and immaterially), and tourists are expected to feel that they have witnessed the essence of traditional village life through the experience, all in one place, and in a way that brings in revenue to the community (Jones, 2009). The motive behind the establishment of cultural villages for tourism is to create opportunities and provide an experience of the rural way of living of local/indigenous people. Cultural villages for tourism showcase the relationship between tourism, heritage, conservation and livelihood activities.

In the context of Botswana, cultural villages for tourism refer to cultural villages that have been established mainly as tourism attractions and for local business and entrepreneurship development. Related to this, the Government of Botswana has established a unit in the Department of Tourism (DOT) to focus on the development of cultural tourism (Saarinen et al., 2014). So far, the division has focused on licensing villages which were more business-oriented, thereby encouraging individually or family owned cultural villages (Monare, 2013). Some cultural villages for tourism have been initiated by communities, created as communally owned 'tourism' enterprises – also known as community-based organisations (CBOs) (see Table 12.2).

CBOs are institutions which have a local area or village theme and tend to be mainly owned and managed by members of a community or one or more villages in a region (GoB, 2007; Rozemeijer, 2001). Many of the cultural villages have been formed under the notion of community-based natural resource management (CBNRM) and are owned as CBOs or trusts. CBNRM has become a popular policy tool in Botswana and southern Africa, in general, highlighting the role of local communities and people in natural and cultural resource management (see Moswete et al., 2009a; Nelson & Agrawal, 2008). As a strategy, CBNRM states that local communities should have direct control over the utilisation of and benefits obtained from their natural and cultural resources (Child, 2004; GoB, 2007); and by securing the control and benefits thereof, local people are also assumed to value and manage them in a sustainable way (Blaikie, 2006; Mutandwa & Gadzirayi, 2007).

Cultural villages for tourism that have been created under the CBNRM model include Xaixai cultural village (Okavango Delta) (Mbaiwa & Sakuze, 2009); Domboshaba cultural village (Kalakamati) and Shandereka cultural village (Sankuyo) (Ford et al., 2010). The tourist activities offered have contributed towards creation of employment opportunities for rural

Table 12.2 Cultural villages for tourism in Botswana

Name	Place (village)	CBO*Formation	Private
Bahurutshe Cultural Village	Mmankgodi	–	Yes
Motse Cultural Village	Kanye	–	Yes
Basiamisi Cultural Village	Serowe	–	Yes
Ditlhakane Cultural Village	Kumakwane	–	Yes
Khudu Cultural Village	Molepolole	–	Yes
Sexaxa Cultural Village	Maun	Yes	–
Domboshaba Cultural Village	Kalakamati	Yes	–
Shandereka Cultural Village	Sankuyo	Yes	–

Source: Department of Tourism (2014).

communities. Some individuals make and sell handicrafts to visitors and tourists who come to the area (Ford et al., 2010).

Botswana cultural villages: Case examples

Bahurutshe and Motse cultural villages for tourism are located in the southern part of Botswana (Figure 12.1). Two cultural villages are located in areas where tourism development is at an early stage (see Butler, 1980). The villages or 'cultural lodges' as they are commonly called in Botswana were established mainly to showcase and protect and preserve the cultures of the Bahurutshe and Bangwaketse, respectively. Both villages are privately owned

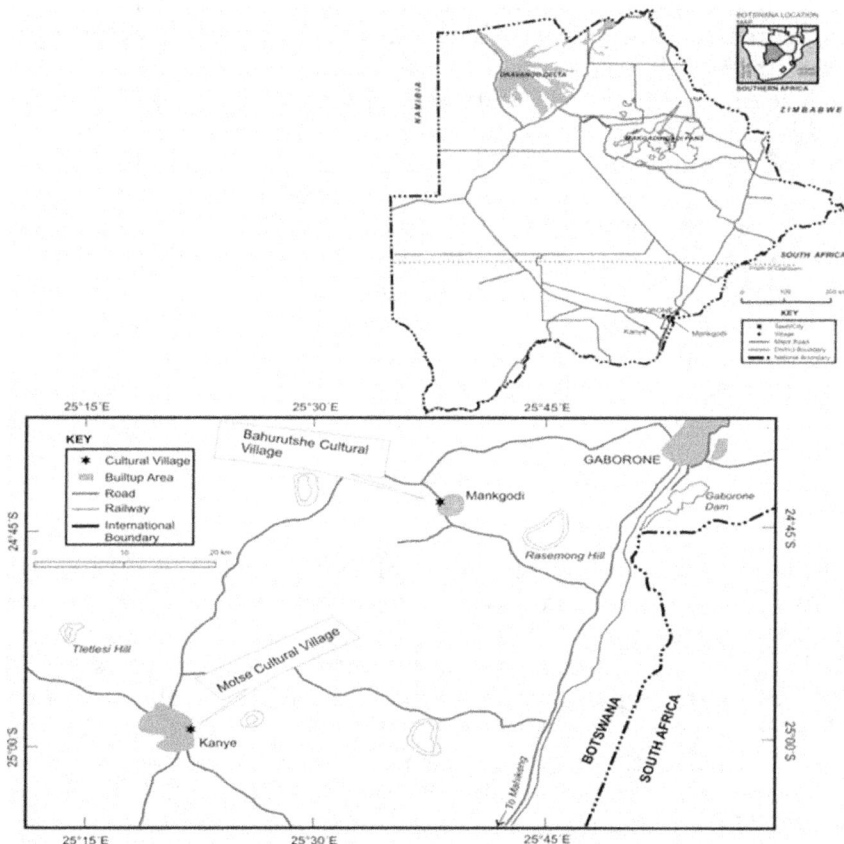

Figure 12.1 Map of Botswana showing study site geographical location (created by P.G. Koorutwe).

and family-oriented tourism enterprises. The management, marketing and activities offered are all in the hands of the owners. Most of the planning as well as the running of the activities is vested in the hands of the individual owners.

Bahurutshe cultural village

Bahurutshe cultural village is located in Mmankgodi which is a rural village in the Kweneng district (see Figure 12.1). The Bahurutshe cultural village was established in 2005, and was mainly set up to showcase the culture of the Bahurutshe, and also as a 'tourism business' (Enviro GIS Consultants, 2004). The Bahurutshe people are one of the ethnic groupings found in southern Botswana. They originated from Lehurutshe in the North West Province of South Africa, and they arrived in Botswana around the 1800s when they were fleeing from Afrikaners' attacks or tribal clashes (Monare, 2013).

The site on which the Bahurutshe cultural village for tourism is situated is a reconstructed Hurutshe village, offering accommodation in the form of traditional rondavels (Plate 12.2) (see Grant & Grant, 1995). It was established to preserve and celebrate the culture of the Bahurutshe and also as tourism commerce. It offers arts, crafts, traditional dancing, educational cultural tours as well as a variety of traditional activities (see Keitumetse *et al.*, 2007). These include, amongst others: a traditional African wedding, traditional dancing, storytelling, traditional food, cuisine tasting, and traditional games such as *mhele, morabaraba* (game of stones). Traditional meals include: sorghum, maize meal, pounded meat, wild spinach, mophane worm, wild beans, as well as traditional beer (Monare, 2013).

Motse cultural village

The Motse cultural village for tourism is located in Kanye in Ngwaketse district. Kanye was originally called Bangwaketse village; Bangwaketse is one of the largest tribes (ethnic groups) in Botswana. History states that the Bangwaketse came to Botswana 400 years ago and settled a short distance east of the capital city of Botswana – Gaborone (GoB, 2001). After fighting a series of battles with other tribes due to strained relations, they moved further south near the Molopo river. Around 1800, they entered the Kanye area and built a stonewalled village on top of Kanye hill, the remains of some of the walls of which are still visible (GoB, 2001, 2003).

Motse cultural village was established in 2004. The main reason for the creation of the cultural village was to establish a commercial entity that

Plate 12.2 Bahurutshe cultural village in Mmankgodi, Botswana. The village provides accommodation and cultural and touristic elements, including 'edutainment', integrating entertainment with cultural education aspects.
Source: Jarkko Saarinen.

would draw visitors to experience Bangwaketse culture (Monare, 2013). This would be achieved through the cooking and tasting of traditional foods (melon porridge, cooked dry bean leaves relish, traditional sorghum made beer). Within the village compound, some dwellings are thatched and also decorated with red earth and the display of traditional ornaments such as beer and water storage clay pots is common. The Tswana mud huts were constructed in order to offer a tribute to the artistry and craftsmanship of the Bangwaketse people and revive their history (see Grant & Grant, 1995; Monare, 2013). Tourism-related activities and services offered in Motse cultural village include guided tours, and parades or shows where men and women wear traditional Bangwaketse attire and perform traditional music and dance. There is a beadwork and pottery-making presentation as well as a traditional beer-making demonstration and tasting. Visitors can choose to be served a variety of traditional food and refreshments. Other activities offered at Motse cultural village include consultations with the village traditional healer.

Discussions and Conclusions

The emergence of cultural villages for tourism has become a 'new vehicle' for rural tourism development in Botswana. Achieving accelerated economic growth through tourism could be ensured by diversifying nature tourism with culture-based tourism, and there are already signs of success in the country (see BTDP, 2000; Lenao, 2013; Moswete & Lacey, 2014). Cultural heritage tourism is important as it has the potential to revive cultures (dance and music, local foods, traditional religions) – the realms of development in Botswana. Botswana is endowed with a variety of cultural attractions that include the World Heritage Site of Tsodilo Hills and Museum, Lesoma Battlefield, Lekhubu Island, and Manyana rock paintings; heritage and memorial sites could be explored for tourism advancement. The draft tourism policy review of 2009 centred on community involvement and participation in culture-based tourism.

In the last few decades Botswana's tourism has been more wildlife-based (Moswete & Dube, 2013). Thus, the advent of the Botswana national ecotourism strategy (2003) and CBNRM policy of 2007 have promoted and encouraged direct involvement of citizens in the production of arts and crafts for tourism. Although there is still low awareness of tourism as a business in most parts of the country (Moswete *et al.*, 2009b), the CBNRM policy has emphasised the need to safeguard cultural heritage resources from which citizens should participate and benefit (Saarinen *et al.*, 2014). The Bahurutshe and Motse cultural villages have shown the potential to contribute by creating employment opportunities, and protecting and preserving the tangible and intangible cultures of the two ethnic groups. Also important is that the two cultural villages provide opportunities for leisure and active outdoor recreation in the form of music and dance, and heritage walks in and around the villages. For job opportunities, most of the employees are residents of each area. The majority of the Bahurutshe employees are elderly men and women. At Motse cultural village, workers are of different ages: young and old people. Motse cultural village has recently become popular for hosting festivals and local events (Monare, 2013).

It is worth noting that during the initial development of the two cultural villages for tourism, environmental impact assessment (EIA) studies were conducted. Some of the findings of the EIAs revealed that cultural villages will promote protection and preservation of the cultures in the two communities (Enviro GIS Consultants, 2004). Indeed, the two cultural villages for tourism have contributed positively to the conservation and safeguarding of the culture of the Bahurutshe and Bangwaketse. The buildings of both cultural villages for tourism are constructed, painted and decorated to reflect the culture of the Bahurutshe and Bangwaketse respectively. The huts are circular in shape and are constructed using cow dung and thatched with grass – displaying typical Tswana architecture. The structural design has since been altered by adopting modernised designs to cater for the needs and wishes of the tourists. Some buildings have now been constructed using bricks and mortar instead of a mixture of mud and cow dung as in the Tswana traditions.

The activities offered by both villages include but are not limited to traditional music and dance with troupes from Mmankgodi (Bahurutshe), Kanye (Motse) and the surrounding villages. Other leisure-related activities consist of staged traditional doctor bone throwing, a traditional wedding, playing *diketo* (game of stone), *dikhwaere* (traditional choirs) and heritage walks. These activities are meant to attract visitors and tourists alike but they also help residents to socialise, interact with visitors and develop pride in their own culture.

Hence, we conclude that cultural villages for tourism that are initiated through the CBNRM model should be encouraged to ensure benefits are equally distributed among citizens, especially local people (see GoB, 2007). More opportunities should be created for local communities such that they experience direct tangible benefits. This could reduce citizens' dependence on government support programmes (see Moswete *et al.*, 2009a; Rogerson, 2006) and thus increase empowerment of individuals through cultural heritage tourism as a business (see Moswete & Lacey, 2014; Saarinen & Lenao, 2014). Through CBOs, local people 'own' the resources since they are 'direct' beneficiaries and hence tend to ensure proper management (Lenao, 2013). The requirements for establishing cultural villages for tourism should not be too demanding as they could end up discouraging citizens from establishing culture-based organisations. CBOs or trusts that promote cultural villages for tourism are a necessity as they will encourage rural communities to work together towards a common goal. This promotes shared culture in a community as well as giving people a sense of pride and belonging (see Rogerson, 2006; Saarinen *et al.*, 2014).

As cultural tourism is relatively new in Botswana, there are many unknown and under-developed cultural heritage sites that could be explored for rural and/or heritage tourism. Accordingly, these cultural villages are private and family owned enterprises, there is low involvement and minimal benefits for the residents of both Mmankgodi and Kanye. In this light we recommend the introduction of community-based cultural tourism that will be managed by and the benefits be shared among members of a community.

References

Bachleitner, R. and Zins, A. (1999) Cultural tourism in rural communities in the residents' perspective. *Journal of Business Research* 44, 199–209.

Blaikie, P. (2006) Is small really beautiful? Community-based natural resource management in Malawi and Botswana. *World Development* 34 (11), 1942–1957.

Botswana Tourism Development Programme (BTDP) (2000) *The Existing and Potential Tourist Attractions in Botswana*. Gaborone: Ministry of Commerce and Industry, Republic of Botswana.

Butler, R.W. (1980) The concept of a tourism area's cycle of evaluation: Implementations for management of resources. *Canadian Geographer* 24 (1), 5–12.

Child, B. (ed.) (2004) *Parks in Transition: Biodiversity, Rural Development and the Bottom Line*. London: Earthscan.

Cole, C. (2007) Beyond authenticity and commoditization. *Annals of Tourism Research* 34 (4), 943–960.

Enviro GIS Consultants (2004) *Bahurutshe Cultural Lodge. Preliminary Environmental Impact Assessment (Final Report)*. Gaborone: Botswana.

Ferrar, T. (1995) Makgadikgadi/Nxai Pan Management Plan. Final Draft. Prepared on behalf of the Department of Wildlife and National Parks as part of the NRMP USAID Project No: 690-0251.33. Gaborone: Under the auspices of the IUCN.

Ford, R., Mlambo, A. and Molwane, A. (2010) *Exploring Social Studies*. Gaborone: Heinemann Pearson.

Gordon, R. (1990) The prospects for anthropological tourism in Bushmanland. *Cultural Survival Quarterly* 14 (1), 6–8.

Government of Botswana (GoB) (2001) *Botswana National Atlas. Department of Surveys and Mapping, Botswana*. Gaborone: Government Printer.

Government of Botswana (GoB) (2003) *The National Development Plan Nine (NDP 9). NDP 9 for 2003/2004 to 2008/2009. Ministry of Finance and Development Planning*. Gaborone: Government Printer.

Government of Botswana (GoB) (2007) Community-based natural resources management policy (*Government Paper No. 2, Ministry of Environment, Wildlife and Tourism*). Gaborone: Government Printer.

Grant, S. and Grant, E. (1995) *Decorated Homes in Botswana*. Phuthadikobo Museum. Gaborone: Bay Publishing.

Gurung, D. (1995) Tourism and gender: Impact and implications of tourism on Nepalese women. A case study from the Annapurna Conservation Area Project (*Mountain Enterprises and Infrastructure Discussion Paper, 95/03*). Kathmandu: International Centre for Integrated Mountain Development.

Ivanovic, M. (2008) *Cultural Tourism*. Cape Town: Juta.

Jones, R. (2009) Cultural tourism in Botswana and the Sexaxa cultural village: A case study (*ISP Collection, Paper 725*). See http://digitalcollections.sit.edu/isp_collection/725

Keitumetse, S. (2005) Sustainable development and archaeological heritage management: Local participation and monument tourism in Botswana. PhD thesis, Department of Archaeology, University of Cambridge.

Keitumetse, S., Matlapeng, O. and Monamo, L. (2007) Cultural landscapes, communities and world heritage: In pursuit of the local in the Tsodilo Hills, Botswana. In D. Hicks, L. McAtackney and G. Fairclough (eds) *Envisioning Landscape* (pp. 101–119). Washington, DC: Leaf Cost Press Inc.

Lenao, M. (2013) Challenges facing community based cultural tourism development at Lekhubu Island, Botswana: A comparative analysis. *Current Issues in Tourism* 16, 1–16.

MacDonald, R. and Jolliffe, L. (2003) Cultural rural tourism: Evidence from Canada. *Annals of Tourism Research* 30 (2), 307–322.

Mbaiwa, J.E. (2005) The socio-cultural impacts of tourism development in the Okavango Delta, Botswana. *Journal of Tourism and Cultural Change* 2 (3), 163–185.

Mbaiwa, J.E. (2011) Cultural commodification and tourism: The Goo-Moremi Community, Central Botswana. *Tijdschrift voor Economische en Sociale Geografie* 102 (3), 290–301.

Mbaiwa, J.E. and Sakuze, L.K. (2009) Cultural tourism and livelihood diversification: The case of Gcwihaba Caves and XaiXai village in the Okavango Delta, Botswana. *Journal of Tourism and Cultural Change* 7 (1), 61–75.

McKercher, B. and du Cros, H. (2002) *Cultural Tourism: The Partnership between Tourism and Cultural Heritage*. Binghamton: The Haworth Press.

Monare, M. (2013) Local people's attitudes and perceptions towards benefits and costs of cultural tourism: Case studies of Bahurutshe and Motse cultural villages, Botswana. MA thesis, University of Botswana, Gaborone.

Moswete, N. and Dube, O.P. (2013) Wildlife-based tourism and climate: Potential opportunities and challenges for Botswana. In L. D'Amore and P. Kalifungwa (eds) *Meeting the Challenges of Climate Change to Tourism: Case Studies of Best Practice* (pp. 395–416). Cambridge: Cambridge Scholars Publications.

Moswete, N. and Lacey, G. (2014) Women cannot lead: Empowering women through cultural tourism in Botswana. *Journal of Sustainable Tourism* 23 (4), 600–617.

Moswete, N., Thapa, B. and Lacey, G. (2009a) Village-based tourism and community participation: A case study of the Matsheng villages in southwest Botswana. In J. Saarinen, F. Becker, H. Manwa and D. Wilson (eds) *Sustainable Tourism in Southern Africa: Local Communities and Natural Resources in Transition* (pp. 189–209). Bristol: Channel View Publications.

Moswete, N., Toteng, E.N. and Mbaiwa, J. (2009b) Resident involvement and participation in urban tourism development: A comparative study in Maun and Gaborone, Botswana. *Urban Forum* 19 (4), 381–394.

Mutandwa, E. and Gadzirayi, C.T. (2007) Impact of community-based approaches to wildlife management: Case study of the CAMPFIRE programme in Zimbabwe. *International Journal of Sustainable Development and World Ecology* 14, 336–344.

Nelson, F. and Agrawal, A. (2008) Patronage or participation? Community-based natural resource management reform in sub-Saharan Africa. *Development and Change* 39 (4), 557–585.

Richards, G. (1997) The social context of cultural tourism. In G. Richards (ed.) *Cultural Tourism in Europe* (pp. 39–52). Wallingford: CAB International.

Richards, G. (2003) What is cultural tourism? In A. van Maaren (ed.) *Erfgoed Voor Toerisme. Nationaal Contact Monumenten*. See www.docstoc.com

Rozemeijer, N. (ed.) (2001) *Community Based Tourism in Botswana: The SNV Experience in Three Community Tourism Projects (Case Studies in /Xai-/Xai, D'kar and Ukhwi)*. Gaborone: SNV Publication (Netherlands Development Organisation).

Rogerson, C.M. (2006) Pro-poor local economic development in South Africa: The role of pro-poor tourism. *Local Environment* 11, 37–60.

Saarinen, J. (2009) Conclusion and critical issues in tourism and sustainability in southern Africa. In J. Saarinen, F. Becker, H. Manwa and D. Wilson (eds) *Sustainable*

Tourism in Southern Africa: Local Communities and Natural Resources in Transition (pp. 269–286). Bristol: Channel View Publications.

Saarinen, J. (2010) Local tourism awareness: Community views in Katutura and King Nehale Conservancy, Namibia. *Development Southern Africa* 27 (5), 713–724.

Saarinen, J. (2011) Tourism, indigenous people and the challenges of development: The representations of Ovahimbas in tourism promotion and community perceptions towards tourism. *Tourism Analysis* 16 (1), 31–42.

Saarinen, J. and Lenao, M. (2014) Integrating tourism to rural development and planning in the developing countries. *Development Southern Africa* 31 (3), 363–372.

Saarinen, J., Moswete, N. and Monare, M.J. (2014) Cultural tourism: New opportunities for diversifying the tourism industry in Botswana. *Bulletin of Geography. Socio-Economic Series* 26, 7–18.

Smith, W. (2009) To infinity and beyond. In D. Bell and M. Parker (eds) *Space Travel and Culture: From Apollo to Space* (pp. 204–212). Oxford: Blackwell.

Stephen, L. (1992) Marketing ethnicity: Zapotec women in Mexico have won – and lost – from the popularity of local crafts in stores up north. In Ecumenical Coalition on Third World Tourism (ECTWT), Tourism and Indigenous People: A Resource Guide. *Contours* 5 (2), 86.

United Nations World Tourism Organisation (UNWTO) (2008) *Policy for the Growth and Development of Tourism in Botswana. UNWTO/Government of Botswana Project for the Formulation of a Tourism Policy for Botswana, July 2008*. Gaborone: UNWTO and Department of Tourism.

Weiman, E. (1993) Tourism: Preservation or perversion of aboriginal culture in Taiwan. In Ecumenical Coalition on Third World Tourism (ECTWT), Tourism and Indigenous People: A Resource Guide. *Contours* 6 (2), 48–53.

World Travel and Tourism Council (WTTC) (2007) *Botswana: The Impact of Travel and Tourism on Jobs and the Economy*. London: World Travel and Tourism Council.

World Travel and Tourism Council (WTTC) (2013) *The Economic Impact of Travel and Tourism 2013*. London: World Travel and Tourism Council.

13 Cultural Tourism in Southern Africa: Progress, Opportunities and Challenges

Naomi Moswete, Jarkko Saarinen and Haretsebe Manwa

Introduction

Cultural resources have become significantly important as tourist attractions in southern Africa. The region has diverse cultures and heritage that include traditional music and dances of various ethnicities; gastronomy; traditional architecture, language, dress, traditional African religions, traditional and modern agricultural practices. Hence, this book attempted to focus on and holistically present cultural tourism in the southern African region using case studies as well as issues pertinent to the tourism industry and its development. Generally, the book aimed to identify and elucidate issues relatable to cultural heritage attractions, tourism products diversification; sustainability and ethical issues of the product; host–guest relationships, level of development and local people's participation in and benefit from tourism. In all, the book deliberates on responsibility and sensitivity in the utilisation, management and marketing of cultural resources for tourism, especially in relation to local communities living in rural and peripheral environments.

In respect to the management and research of cultural tourism there is a need to understand communities and their specific cultural and natural resource values and utilisation priorities. This calls for deeper understanding of what community means as an idea beyond a 'spatial location' alone. As demonstrated in the book, community characteristics such as ethnicity, gender and livelihood composition are connected with issues of power, empowerment, participation and inequality in (cultural) tourism development. Understanding communities and their specific characteristics, history and internal and external dynamics has an academic value but also great applied value when thinking about local development impacts or sustainability in cultural tourism development and management.

With reference to the theme of this book, and perspectives from different case studies, it can be concluded that indeed cultural tourism is a complex, broad-based and constantly transforming phenomenon. This general conclusion is affirmed by many previously outlined definitions and typologies (see McKercher & du Cros, 2002; Richards, 1997, 2003; Van der Ark & Richards, 2006) indicating that cultural tourism is:

- A form of special interest tourism where culture forms the basis of either attracting tourists or motivating people to travel to specific destinations and/or sites.
- An experiential activity where engagement with culture can be unique and intense, which often involves educational activities and sites where visitors can also be entertained.
- A business perspective involving the development and marketing of various cultural tourism sites and attractions where tourists participate in a large array of activities or experiences.

In addition, cultural tourism can be defined from community and host perspectives as a tool for cultural revitalisation, heritage conservation, local development, economic diversification and poverty reduction (see Grünewald, 2002; Saarinen et al., 2011; Smith & Richards, 2013; Monare et al. in Chapter 12 also confirm this in the case studies of Botswana).

Cultural tourism has become a niche and lucrative form of tourism in southern Africa. It has proved of benefit in terms of conservation and preservation of unique resources that include the Tsodilo Hills World Heritage Site (see Mbaiwa, Chapter 8) and Lekhubu Island (see Lenao, Chapter 10) in Botswana, Bushman/San painting in the Brandberg National Monument area (see Lapeyre, Chapter 11) in Namibia and iSimangaliso Wetland Park (see Ndlovu, Chapter 4) in South Africa, for example.

Diversification and Cultural Tourism Products

The southern African region is renowned as a cultural heritage resource rich destination that spans from local/community scale cultural sites to

UNESCO World Heritage Sites (e.g. Robben Island, South Africa; Great Zimbabwe National Monument in Zimbabwe; and Twyfelfontein, Namibia) depicting the region as having diverse cultures, histories, memorial sites, museums, art galleries, old mine sites, caves, ethnic groupings and many more that attract visitors and tourists from afar who tend to be motivated and attracted by the uniqueness and diversity of the southern African region. Based on the existing literature (see Ivanovic, 2008; McKercher & du Cros, 2002; Richards, 2003) and different chapters of this book a typology of cultural heritage-related resources that abound the region can be listed in a following way:

- Heritage sites (archaeological sites, whole towns, monuments, museums).
- Performing arts venues (theatres, concert halls, cultural centres).
- Visual arts (galleries, sculpture parks, photography museums, architecture).
- Festival and special events (arts festivals, sporting events, carnivals).
- Religious sites (cathedrals, temples, pilgrimage destinations, spiritual retreats).
- Rural environments (village farms, national parks, eco museums).
- Indigenous communities and traditions (tribal people, ethnic groups, minority cultures).
- Arts and crafts (textile, pottery, painting, sculpture).
- Language (learning or practice).
- Gastronomy (wine testing, food sampling).
- Industry and commerce (factory visits, mines).
- Modern popular culture (pop music, shopping, fashion, media design, technology).
- Creative activities (painting, photography, dance).

While a culturally rich and diverse region, the tourism industry in southern African has been largely focused on wildlife and wilderness resources that include 'The Big 5' (elephant, buffalo, leopard, lion and rhinoceros). These wildlife and wilderness resources are used largely in marketing programmes (e.g. see RETOSA: www.retosa.co.za/). Indeed, promotional and marketing narratives and glossy magazines showcasing charismatic species of wild animals that include but are not limited to elephants, buffalo and the big cats are prolific across Botswana, Namibia, South Africa, Zambia and Zimbabwe, for example. Many long-haul travellers who frequent the region come specifically to see and experience wild animals and their related habitats that incorporate vast and unique natural landscapes of the kingdom of Lesotho, the

Kalahari and Namib deserts, and the expanse of Victoria Falls that sits astride Zambia and Zimbabwe. Therefore, many countries like Botswana and Namibia have been sluggish in identifying and integrating cultural attractions as important elements of their tourism industry (see Monare et al., Chapter 12). However, recently many of the southern African countries have started to recognise cultural tourism as an add-on product to nature and wilderness attractions (see Monaheng, Chapter 3).

During the past decade there has been a growth in the number of regional and international visitors to the region, hence competition has increased, compelling countries to diversify their product base from safari and wildlife-based tourism to incorporate cultural activities. New attractions and products such as cultural villages for tourism, open-air and living museums, carnivals and arts festivals (see Monare et al., Chapter 12; Njerekai, Chapter 7; Pretorius, Chapter 6) have emerged as activities and sites that attract a growing number of visitors. However, so far this increasing cultural consumption has mainly been based on overseas visitors who want to visit southern Africa to indulge in cultures totally different from theirs. While a positive trend, this will also warrant careful and ethical packaging and marketing of sensitive and fragile cultural sites to avoid irrevocable damage resulting from tourist activities (see Mbaiwa, Chapter 8; Lapeyre, Chapter 11; and Lenao, Chapter 10).

Development and Local Benefits

Local communities have begun to reap benefits from culture-based tourism. Accordingly, Lapeyre (Chapter 11) uncovers that the evolving tourism industry has provided possibilities for local households to improve their livelihoods in a region where very limited alternative options than tourism exist. Similarly, Saarinen (Chapter 2) states that cultural tourism can provide development for local communities; however, it is imperative that sustainable tourism development approaches are considered particularly when integrating tourism and local ethnic, indigenous communities and indigenous knowledge systems (see also Monaheng, Chapter 3). Integrating these can reduce issues of conflict between all who have a stake in the region's resource base: women, indigenous people, heritage resource managers, and traditional practitioners of agriculture, visitors and the private sector, for example.

Other benefits are demonstrated where there is safeguarding of cultures and heritage resources. This is evident in the case of new tourism attractions such as Bahurutshe cultural village (Botswana); Shakaland cultural village (South Africa), and Damaraland Living Museum (Namibia). Or in situations where communities have formed community-based organisations (trusts) or conservancies (see Kimaro and Nengola, Box 11.1; and Lenao, Chapter 10). These institutions have started to provide benefits in terms of preserving cultures. Opportunities for employment are evident and local people and the international community find solace and peace, social capital, leisure and recreation (cultural exchange). In addition, communities are able to show what MacCannell (1976) calls front stage where they portray what they believe tourists would like to see – a staged authenticity of the Shona culture (see Mamimine & Madzikatire, Chapter 9).

Cultural Tourism, Poverty Alleviation and Rural Development

In southern Africa the history of cultural tourism is relatively short. Despite being an evolving form of tourism it is increasingly supported and favoured as it can bring financial returns to individuals and businesses without major capital investments (see Saarinen, Box 12.1) and during the last two decades southern African governments have identified and earmarked cultural tourism as a strategy for the upliftment of the rural poor via alleviating extreme poverty in rural and/or remote areas (see Lapeyre, Chapter 11; Lenao, Chapter 10; Monaheng, Chapter 3). Subsequently, culture-based tourism has also become beneficial as it can play a vital role in rural development (see Monaheng, Chapter 3; Njerekai, Chapter 7; Saarinen, Chapter 2). Therefore, identification of cultural resources – both tangible and intangible ones – became topical as the rapid growth of international tourism was already negatively affecting the natural resources in some southern African countries where tourism is more advanced. As is evident, however, human activities are also rapidly altering cultural landscapes and heritage resources (defacing of monuments, over collection of cultural items/goods for survival), especially in areas where poverty is/was rampant and traditional livelihood opportunities have dwindled (see Lenao, Chapter 10; Mbaiwa, Chapter 8).

In this context, cultural tourism has a potential to reduce poverty in many regions and communities in southern Africa (e.g. Lesotho, see Monaheng, Chapter 3; Botswana, see Monare et al., Chapter 12; Zimbabwe, see Njerekai, Chapter 7; and Namibia, see Lapeyre, Chapter 11). Indeed, cultural tourism can contribute towards the creation of employment opportunities and thereby reduce rural–urban migration as individuals and families venture into craft-related projects (see Lapeyre, Chapter 11; Manwa, Box 8.1; Monaheng, Chapter 3). Uncontrolled cultural tourism, however, has negative repercussions; for instance, cultural commodification (where traditional or indigenous items face the threat of losing cultural values as they are exposed and sold to the tourists) of Tsodilo Hills (Mbaiwa, Chapter 8), or the use of the Ovahimbas in tourism promotion as a posing object for tourists to gape at and consume (see Saarinen, Chapter 2).

Sustainability and Cultural Tourism

Tourism in cultural areas has a range of impacts; some less beneficial or more destructive than others (see Mbaiwa, Chapter 8; and Laperye, Chapter 11). It is evident that as the tourism industry grows, visitations to cultural sites equally increase – hence, the need for adaptive management techniques may be imperative to curb changes in the natural cultural heritage landscape. Sustainability implies a state of equilibrium in which the activities of the human population coexist in broad harmony with their natural, social and cultural limits (see Lenao, Chapter 10; Laperye, Chapter 11; and Mamimine and Madzikire, Chapter 9). Sustainable tourism is envisaged as leading to the management of cultural resources in such a way that social, economic and aesthetic needs can be fulfilled while maintaining cultural integrity, essential ecological processes, biological biodiversity and life support systems (Middleton & Hawkins, 1998: 247).

Ethical Issues in Cultural Tourism

Due to globalisation and development of transportation and travel technologies, the rapid increase in the demand for cultural tourism experiences is one of the major challenges that heritage resource managers face worldwide, including in southern Africa (Breen, 2007). Many governments and regional actors have also responded from below to the new possibilities globalisation potentially offers to growth and development. Indeed, governmental bodies in the region have realised the opportunity and need to develop tourism for economic diversification; revitalise dying/decaying cultural heritage resources; create jobs for citizens; protect and conserve cultural attributes and values of landscapes and historical structures. There have been debates on how to package, expose and share cultural values carefully to the tourism market without detrimentally devaluing cultures. However, the receipt of quick tourist dollars experienced by cultural tourism enterprises and operators has led to the challenges of non-observation and maintaining respect for other cultures or citizen minorities (see Hitchcock & Brandenburgh, 1992; Saarinen, Chapter 2). With the advent of new technologies and deepening globalisation, cultural tourists will be in search of sites and monuments which convey an aura of the past and traditions, and specialised skills in the interpretation of unique cultures (Salazar, 2005). There is a need to devise proper policies that will ensure sustainable utilisation of cultural resources, respect for cultures (beliefs, sacredness of sites and places) and desist from using citizen minorities for marketing and promotion of tourism.

Thus, in spite of the many evident positive impacts of cultural tourism in southern Africa, there is an urgent need for all stakeholders to collaborate in order to reduce loss of human dignity and cultural destruction. Native residents need to be respected and allowed to have a voice in all spheres of the tourism sector (see Mbaiwa, Chapter 8; Saarinen et al., 2011). They also need to have a right to say no to tourism and commodification of their culture if they so wish.

Participation of Local People

International tourists who visit the region are generally interested in seeing and learning about local cultures, such as the San and Khoikhoi or other ethnic groups (see Mamimine & Madzikire, Chapter 9; Monare et al., Chapter 12; Saarinen, Chapter 2). Thus, in order for cultural tourism to grow and for the sustainable management and utilisation of cultural resources there is a need to actively involve rural and local people in decision-making and management to reduce and avoid conflicts (see Lenao, Chapter 10; Monaheng, Chapter 3).

Educating stakeholders, including local people and tourists, about cultural heritage resources for sustainable utilisation and development of attractions and destinations for tourism is of paramount importance for the southern African region. As alluded to in most of the chapters in this book (see Manwa, Box 8.1; Saarinen, Box 12.1; Lenao, Chapter 10; Mbaiwa, Chapter 8; Monaheng, Chapter 3; and Monare et al., Chapter 12), education is regarded as the answer to the problems and challenges facing the advancement of the culture-based tourism industry in the region. In most parts, both formal and informal education have become some of the key mechanisms in minimising the (local) heritage site management impacts of tourists/visitors and local/rural communities: e.g. in the preservation of the culture of the Bahurutshe in Botswana (see Monare et al., Chapter 12); the authenticity of the Basotho hat (see Manwa, Box 8.1); conservation of aesthetic beauty, naturalness, pristine and human-made cultural heritage sites (see Lenao, Chapter 10).

Another important stakeholder in sustainable cultural tourism is the non-governmental organisations (NGOs). NGOs have played an important role in building human capital and advocacy work. They have been able to guard against exploitation of people by the tourism industry as well as advancing people's voice against development of undesirable forms of tourism (Barnett, 2008). In addition to advocacy work, NGOs also become tourists – as volunteer tourists. Volunteer tourism is now advocated as an alternative to ordinary cultural tourism. There is a deeper and more meaningful cultural exchange between volunteers and hosts in comparison to the superficial experience of an ordinary cultural tourist (McIntosh & Zahra, 2007).

A key issue that the book has demonstrated is that although South Africa dominates the international and domestic markets in cultural tourism or tourism in general, the wider southern African regional tourism market is growing, rich and provides cultural tourism attractions and sites that a single country cannot easily match. In order to capitalise on this regional competitiveness, a deeper and mutual collaboration between countries, regions and key stakeholders is needed. However, it should be noted with caution that each southern African country has unique tourism and regional development contexts and related challenges in diversifying the tourism product, for example. However, many of these challenges can seemingly be ameliorated and rectified via better communication, networking and public education. Thus, southern African regional governments should put more emphasis on glocalisation, whereby the needs and wants of tourists are assimilated into the local cultures as in dance, music and souvenir production (Mamimine & Madzikatire, Chapter 9). In addition, southern African countries should also pursue cross-border collaboration and educational empowerment of citizens in remote areas, urban communities (especially those living in historic towns and villages) and other stakeholders about the need to diversify the tourism resource base of the entire region. This would also expand the 'regional tourism product' away from a wildlife and wilderness resource-dominated tourism landscape towards cultures and heritage resources that are found in abundance in the region.

Obviously there are challenges to utilising and harvesting the regional cultural resources in tourism. The different case studies in this book have aimed to critically but constructively uncover that (cultural) tourism industry and related promoters have a general tendency of commercialising host cultures in their operations. It is clear that the industry can commoditise, deodorise and sanitise cultural resources and values for the consumption of tourists (see Jafari, 1996: 45). Thus, there is a need for responsible marketing and sustainable utilisation and governance of cultural resources in tourism (see Hitchcock & Brandenburgh, 1992; Middleton & Hawkins, 1998). Keeping the balance between the development of cultural tourism and local cultural values and needs in southern Africa is of paramount importance. In addition, the capacity and tourism awareness of local people are not always at levels that empower their full and beneficial participation in tourism development. Overall, it is evident that education and responsible production and consumption of sensitive and fragile cultural heritage resources have the potential to contribute positively to the understanding and use of culture in tourism and leads towards conservation and preservation of cultural resources also in the future.

References

Barnett, T. (2008) Influencing tourism at the grassroots level: The role of NGO tourism concern. *Third World Quarterly* 29 (5), 995–1002. DOI: 10.1080/01436590802106213.
Breen, C. (2007) Advocacy, international development and World Heritage Sites in sub-Saharan Africa. *World Archaeology* 39 (3), 355–370. DOI: 10.1080/00438240701464772.
Grünewald, R.D.A. (2002) Tourism and cultural revival. *Annals of Tourism Research* 29 (4), 1004–1021.
Hitchcock, R.K. and Brandenburgh, R.L. (1992) Tourism, conservation and culture in the Kalahari Desert, Botswana. Tourism and indigenous people: A resource guide. Ecumnical Coalition on Third World. *Contours*, 39–42.
Ivanovic, M. (2008) *Cultural Tourism*. Cape Town: Juta.
Jafari, J. (1996) Tourism and culture: An inquiry into paradoxes. Proceedings of a round table on Culture, Tourism, Development: Crucial Issues for the XXIst Century. Paris: CLT/DEC/SEC-1997.
MacCannell, D. (1976) *The Tourist. A New Theory of the Leisure Class*. New York: Schoken Books.
McIntosh, A.J. and Zahra, A. (2007) A cultural encounter through volunteer tourism: Towards the ideals of sustainable tourism? *Journal of Sustainable Tourism* 15 (5), 541–556. DOI: 10.2167/jost701.0.
McKercher, B. and du Cros, H. (2002) *Cultural Tourism: The Partnership between Tourism and Cultural Heritage*. Binghamton: The Haworth Press.
Middleton, V.T.C. and Hawkins, R. (1998) *Sustainable Tourism: A Marketing Perspective*. Oxford: Butterworth-Heinemann.
Richards, G. (1997) The social context of cultural tourism. In G. Richards (ed.) *Cultural Tourism in Europe* (pp. 39–52). Wallingford: CAB International.
Richards, G. (2003) What is cultural tourism? In A. Van Maaren (ed.) *National Contact Monument*. See www.docstoc.com (accessed 30 October 2013).
Saarinen, J., Rogerson, C. and Manwa, H. (2011) Tourism and millennium development goals: Tourism for global development? *Current Issues in Tourism* 14 (3), 201–203.
Salazar, N.B. (2005) Tourism and globalisation, 'local' tour guiding. *Annals of Tourism Research* 32 (3), 628–646.
Smith, M. and Richards, G. (eds) (2013) *The Routledge Handbook of Cultural Tourism*. New York; London: Routledge.
Van der Ark, A. and Richards, G. (2006) Attractiveness of cultural activities in European cities: A latent class approach. *Tourism Management* 27, 1408–1413.

Index

accommodation
    campsites 113, 117–118, 138, 152–153
    facilities for business travel 34
    lodges 113, 153–155
    multinational hotel chains 24
    staff quarters at tourist lodges 155, 157
    traditional, in cultural villages 174
Afrikaner cultural heritage 67
agriculture
    challenges for coexistence with tourism 11–12, 138–139
    economic and practical difficulties in Africa 134
    sale of produce to tourists 140–141
    tours to working cattle posts 141–142
apartheid legacy problems, SA 65–67, 70
archaeological sites
    Brandberg Mountain bushman paintings (Namibia) 145, 146
    Lekhubu Island (Botswana) 136
    Tsodilo Hills rock art (Botswana) 107–108, 113
    visitor attractions in Botswana 170, 171
artisans
    'genuine' production for indigenous use 167
    rural–urban trade in traditional products 35–36
arts festivals
    aims and objectives 80–82
    challenges and problems 82
    range and importance in South Africa 10, 80, 83
    see also carnivals
arts tourism, related to culture 78–79
Australian aboriginal culture 7

authenticity
    of the Basotho hat 104–106
    concept, in tourism studies 166–168
    culture preservation and dynamic change 37–38, 43, 103
    risks of arts festival commercialisation 82
    in showcased 'Otherness' for visitors 121, 122, 124–126

Bahurutshe cultural village (Mmankgodi) 173, 174, 175
Bangwaketse (Motse) cultural village, Botswana 173, 174–175
Basotho people (Lesotho)
    cultural heritage resources 38
    as custodians of culture 9, 37–38, 41–42
    festivals and heritage display 40
    traditional dress 41, 104–106
benchmarking research 58, 71–72
'Big Hole' diamond mine, Kimberley (SA) 62–64
Botswana
    CBNRM projects 11, 135
    cultural villages 12–13, 25–26, 171–175
    heritage site management 10–11, 109–111
    important cultural heritage attractions 170–171
    tourism development strategy 132–133
    World Heritage Sites 101, 107
    see also Lekhubu Island; Tsodilo Hills
Brandberg Mountain (Namibia)
    accommodation staff, employment conditions 155–159
    community empowerment and participation 157–159, 162–163

map and tourism resources 145–147
tourist accommodation 12, 152–154
White Lady Lodge partnership contract 153–155
buffer zone management 112, 118, 119
business tourism
    globally important forms 34
    informal sector, sub-Saharan Africa 34–36

Canboulay processions (Trinidad) 88, 89
Caribbean carnival traditions 88
Carne Vale festival (Europe) 87
carnivals
    establishment and objectives, Zimbabwe 89–90
    historical origins 87–89
    impacts on participants 92–96
    planning and funding 91–92, 97
    scope and international roles 86, 92–93
cattle *see* livestock
CBNRM (community based natural resources management)
    application of framework in Botswana 11, 135, 176
    objectives for cultural villages 172–173, 177
Chapungu Dance Group (Zimbabwe)
    dances performed 126–128
    formation and members 123
    role and importance of leader 11, 123–126, 128–129
commercialisation
    cultural heritage programmes 54
    local control of development 147, 188–189
    of traditional cultural products 35–36
commodification
    cultural, definition 102–103
    effects on traditional culture 8, 82, 103
    impacts on management, Tsodilo Hills 10–11, 113–118
    packaging of traditional performances 122
communities, local
    awareness of tourism impacts on culture 41–42

benefits of cultural festivals 81, 82, 97
characteristics and scope of concept 181–182
employment prospects and achievement 155–162
importance of indigenous knowledge 32, 33
community based organizations (CBOs)
    Community Conservation Forums, Lesotho 39–40
    Gaing'O Community Trust, Botswana 136, 139
    Tsiseb conservancy, Namibia 146, 147, 152–154, 163
    Tsodilo Community Trust, Botswana 108, 112, 115
community-based tourism
    joint ventures with private investors 153–155
    objectives 32, 53–54
    ownership of cultural villages, Botswana 172–173, 177
    use of indigenous knowledge systems 40–42
conservation
    cultural 20, 24, 37–38, 114, 176–177
    local livelihoods 111, 118
    nature and environment 33, 39–40, 117–118
    skills development 149–150
    wildlife 135
constructed authenticity 167
conventional mass tourism (CMT) 59, 166, 169
cottage industries 35
crafts *see* handicraft production
creative tourism 78
cultural heritage tourism (CHT) *see* heritage tourism
cultural tourism
    definitions 4, 17–18, 76–78, 133, 182
    scope and impacts 18, 31–32, 182
cultural villages
    Helvi Mpingana Kondombolo (Namibia) 148–151
    Lesedi (South Africa) 24–25
    Mmankgodi and Kanye (Botswana) 12–13, 173–175, 176–177

cultural villages (*Continued*)
  organisation and creation policy in Botswana 171–173
  Shakaland (KwaZulu-Natal, SA) 48, 171
  Thaba-Bosiu (Lesotho) 40
  types and establishment in southern Africa 23–24, 25–26, 147–148
culture
  'differences are beautiful' approach 126
  heritage resource surveys 39, 185
  preservation and change 37–38, 43, 94, 105
  as pull factor for tourists 76, 171
  purpose and products 77–78, 81

dance performances
  carnival workshops, Zimbabwe 94
  Lesedi cultural village, SA 24–25
  traditional sexual and warrior dances 126–128
  training for traditional dance skills 123
  see also Chapungu Dance Group
data collection and records 58, 69, 95
  methods and analysis 108–109, 122, 136–138
De Beers Consolidated Mines, 'Big Hole' Project 62–64
definitions
  arts tourism 78–79
  cultural commodification 102–103
  cultural tourism 4, 17–18, 76–78, 133, 182
  heritage/cultural heritage (CHT) tourism 50
  industrial heritage tourism 61–62
  rural tourism 133–134
demonstration effects 8, 18, 82
Department of Museum and National Monuments (Botswana) 109–111, 113
destination management
  challenges in South Africa 58–59, 68–70
  cultural heritage tourism promotion 53–54, 55, 182
  partnership contracts and compliance 154–155
  plans and information sources 108–109, 118–119
  provision of facilities and activities 113, 117–118

risks and benefits of tourism 103, 113–118, 184–185
success factors for cultural villages 148–151, 172–173
and value of carnivals to Zimbabwe 96–97
Diamond Trust, donor funding 111–112
Doxey's irritation index (irridex) model 5, 12
dress
  Basotho hat 104–106
  carnival costumes 91, 94
  traditional seshoeshoe design, Basotho women 41
  used in portrayal of Otherness 125–126

economics
  benefits from Zimbabwe Carnival 95, 96
  development strategies, role of tourism 26, 31, 59, 134
  financial planning for events 91–92
  households, support networks 160–161
  regional development and heritage tourism 54, 176
  well-being of local communities 19, 33–34, 103
ecotourism 32, 39–40, 111, 136, 176
edible wild plants 33, 42, 108
education
  activities offered at cultural villages 174–175
  costs and access for rural poor 159, 161
  environmental, in schools 42
  importance for sustainable cultural tourism 187
  'living history' tools in cultural tourism 53
  skills transfer in carnivals 94
Education for Sustainable Development (ESD) Programme 42
employment
  conditions and wages, lodge staff 155–159
  conflict and linkages, tourism/farming 139, 140
  hiring and firing procedures 154, 156, 158–159

opportunities of tourism development 142, 159, 176, 185
empowerment of local groups
  Australian aboriginal communities 7
  importance for sustainability 12, 43, 177
  limitations at White Lady Lodge, Namibia 157–159
  marginalised groups and heritage tourism 53–54
  partnership contract commitments 153, 154, 155
  rural communities 135, 139
environmental impact assessments (EIA) 176
environmental protection
  carnival clean-up campaigns 93–94, 95–96
  education in schools 42
  needs due to tourism impacts 117–118
  role of indigenous knowledge 33
ethical issues 186–187
ethnic minority groups
  benefits and risks of tourism development 8–9, 18, 26–27, 169–170
  changing environments and awareness 19–20
  definition, compared with 'indigenous' 20–21
  degree of involvement in cultural villages 24, 174–175
  used in destination marketing 22–23, 187
Europe, history of carnivals 87
existential authenticity 167
'exotic'
  exploitation in marketing 20, 22–23
  interpreted as authenticity 126, 127
  tourist search for 5, 124
experience, tourist *see* visitor experience

festivals
  Basotho culture celebration (Lesotho) 40
  Carne Vale (European pre-Lent feast) 87
  definition and cultural meaning 79–80
  hosting at cultural villages 176
  range of types 80
  see also arts festivals; carnivals
FIFA World Cup, 2010 (SA) 59
focus group discussions (FGDs) 137
food
  cookery demonstration/tasting, cultural villages 174, 175
  gathering and preparation, tourist involvement 34
  market potential of traditional foods 140–141
funding
  costs and budgets, Zimbabwe carnivals 91–92
  social investment by donors, Tsodilo Hills 111–112
  types for cultural villages 24, 150
  WWF/USAID support at Brandberg Mountain 152, 153

Gaing'O Community Trust (GCT) 136, 139
gender imbalance, tourism marketing images 23
globalisation impacts
  cultural 4–5, 17, 52, 121
  economic 52, 186–187
glocalisation 188
Grand Tour (of Europe, historical) 5
guides, training and support 110–111, 115, 146

habitat, history, handicrafts and heritage (4Hs) 21
handicraft production 35–36, 42, 77, 139
  craft markets 168
  product authenticity 104–106, 167
Harare International Carnival (Zimbabwe)
  costs and funding sources 91–92
  potential value and impacts 10, 92–97
  themes and events 90
Helvi Mpingana Kondombolo cultural village 148–151
herbal medicines, traditional 35, 42, 48
heritage tourism (CHT)
  community and tourist transformation potential 51–52

heritage tourism (CHT) (*Continued*)
  definitions 50, 61–62, 78
  impacts of core zone fencing 116, 118
  interpretation of sites 67
high-impact visitor groups 117
history
  heritage resources and routes 48–49, 70–71
  historical re-enactments 49–50
  origins of carnivals 87–89
  past indigenous/Western interactions 125, 127
homestays 34, 41, 49
hosts
  attitudes to tourists 5–6
  cultural meaning of festivals 79–80
  pressure to satisfy tourist expectations 23
household income and spending 160–162

Inanda Heritage Route (SA) 48–49
indigenous knowledge systems (IKS)
  importance in Basotho culture 9, 37–38
  integration with natural resource management 33, 116, 118
  scope and relevance to tourism 32, 33–34, 40–42
indigenous people
  associated tourism products 78, 167, 171
  internationally agreed definition 21
  nomadic tribal clans 145
  support and tourism skills training 112, 114
  tourist perceptions of 'primitive' nature 125–126, 127
  traditional leaders 91, 116–117
industrial heritage tourism 61–64
informal sector business tourism 34–36
infrastructure development 112–113, 141, 146, 154
institutional analysis (Ostrom) 12, 152–155
intellectual property rights 33, 42
international tourists
  carnival participants 90, 92, 93
  opinions of cultural heritage experience, SA 65, 69
  proportion visiting a cultural village (SA) 23
  statistical analysis of motivations 60–61, 69
  as targets of promotional material 54, 184
interviews, key informants 108–109, 114–115, 137
irridex (irritation index) model (Doxey) 5, 12
IUCN, Botswana visitor survey 170

jerusarema dance (Zimbabwe) 126–127

Kanye (Motse) cultural village, Botswana 173, 174–175
Kimberley 'Big Hole' diamond mine (SA) 62–64
Korekore people (Zimbabwe) 125
KwaZulu-Natal province (SA)
  cultural and heritage resources 9, 47–49
  tourist motivations and experience 51–52, 55

labour force *see* employment
language
  diversity in South Africa 66
  as indigenous heritage, preservation 37
  used by Chapungu dance leader 124
Lekhubu Island (Botswana)
  cultural tourism and agriculture conflicts 11–12, 138–139
  location and resources 135–136
Lesedi cultural village (SA) 24–25
Lesotho
  community-based tourism ventures 40–42
  cultural product traders, business travel 36
  development strategies for tourism 38–39, 40
  heritage sites and conservation 39–40
  importance of traditional culture 104
  sustainability of cultural resources 9
  see also Basotho people
Letloa Trust (NGO) 108, 110, 114
Liberation Heritage Route (LHR) 71
livelihoods
  diversification in rural areas 139–141, 184

donor support, Tsodilo Hills 111–112
sustainability in local communities 12, 103, 115, 159–163
value of indigenous knowledge 33–34
livestock 116, 138–139, 141–142, 154
local economic development (LED) 61, 150

Maasai culture (Tanzania) 6
Makgadikgadi salt pans (Botswana) 135, 139, 170–171
Malealea Lodge (Lesotho), ecotourism 41, 42
Maloti Drakensberg Transfrontier Programme (MDTP) 39–40
management *see* destination management
Mandela, Nelson 68
Maori heritage (New Zealand) 37
marketing
  cultural heritage product opportunities 53, 54, 105–106
  effectiveness, for carnival promotion 91
  geographical imbalance, SA products 68–69, 71
  political/social acceptability of ethnic images 22–23
  use of internet 150
  wildlife and wilderness bias 183–184
  word-of-mouth, and destination image 96–97
mass tourism, conventional (CMT) 59, 166, 169
mbende (jerusarema) dance 126–127
medicinal plants 33, 39, 42, 108
MICE tourism (business travel) 34
Mmankgodi (Bahurutshe) cultural village, Botswana 173, 174, 175
Mmatshumu village (Botswana) 135, 136, 141
models
  social impacts of tourism 5, 12
  of traditional culture (cultural villages) 24, 148
Morija Arts and Cultural Festival (Lesotho) 40
Moshoeshoe Day, Lesotho 40
motivations (of tourists) 5
  diversity in Tsodilo Hills visitors 112–113
  purpose of visit analysis, SA 60–61, 65
  'serious indigenous tourists' 22

socio-psychological factors 50–52, 166, 169–170
Motse (Bangwaketse) cultural village, Botswana 173, 174–175
muchongoyu dance (Zimbabwe) 127–128
museums
  celebrating historical events/people 48
  cultural preservation and display 40, 110
  Kimberley Mine Museum, SA 62–64
  open-air 149
  potential and challenges, South Africa 54, 59, 67
music
  African cultural heritage preservation 37
  carnival opportunities, Zimbabwe 94, 95
  Shona cultural traditions 122

NACOBTA (Namibian NGO) 146, 152–153
Namibia
  cultural villages 147–151
  local community involvement in tourism 12, 23, 146–147, 163
  national monuments and parks 146, 150
  see also Brandberg Mountain
National Department of Tourism (NDT), SA 65, 68, 71
national identity
  collective development (nation-building) 60, 65–67, 70
  competitiveness and collaboration 188
  global image enhanced by tourism 59–60
  role of cultural villages 24, 26, 147
  unity promoted by carnivals 90, 93
national parks
  Etosha (Namibia), tourism route links 150
  Lesotho/SA, conservation and ecotourism 39–40
  visitor experience in Botswana 170–171
National Tourism Service Strategy (NTSS), SA 65, 68
native people *see* indigenous people
nature (biodiversity) conservation
  community involvement 39–40
  use of indigenous knowledge 33

Ndau people (Zimbabwe) 127–128
negative impacts of tourism 8
　challenges of Harare International Carnival 94–95
　local community awareness of 41–42
　loss of cultural elements (hunting, San culture) 114–115
'new tourist' ('post-tourist') concept 52, 170
non-governmental organizations (NGOs) 187–188

objective authenticity 167
Okavango Delta (Botswana)
　demonstration effect in local communities 8
　wildlife migration restricted by fencing 118
'Otherness'
　concept definition 121–122
　reinforced by dance leader's presentation 124–126, 128–129
　social construction 11, 122–123, 129
Ovahimba people (Namibia) 23, 25, 186

parades, as element of carnivals 86, 88, 91, 96
　traffic and pollution consequences 94–95
participation
　community leaders in carnival organisation 91
　emotional engagement at heritage sites 51–52
　local communities in conservation 39–40
　stakeholders in cultural tourism 187–189
　tourists in farming activities 142
　tourists in local cultural activities 34
　urban communities, in carnival clean-up 93–94
performing arts 81
policies, tourism
　community involvement emphasis 18, 176
　diversification trends 3–4
　implementation challenges 9–10
　nation-building aims 60, 66–67, 147
　for rural development, Botswana 132–133, 171–173
　strategic planning in Lesotho 38–39
politics, influence on tourism in South Africa 65–67, 70
pony trekking 41
'post-tourist' ('new tourist') concept 52, 170
poverty
　associated with traditional culture 42
　government objectives and strategies 38, 69, 137
　potential of cultural tourism for alleviation 31, 71, 105, 185–186
　socio-economic status improvement 159–163
private sector
　accommodation partnerships 152–155, 162–163
　opportunities in cultural heritage tourism 61
　ownership of cultural villages 24, 172, 173–174, 177
　sponsor funding for carnivals 91–92
products, cultural tourism
　historical re-enactments 49–50
　management of cultural villages 24–25, 148–151, 171–175
　political heritage sites and monuments 67, 68
　range and diversity 7–8, 18, 77–78, 182–184
　traditional and artisan goods 35–36, 104–106, 167

rights of cultural minorities 20
　intellectual property 33, 42
Rio carnival (Brazil) 89
roads, rural impacts 113, 141
Robben Island Museum (SA) 67
routes, cultural heritage 48–49, 70–71
rural development
　benefits of cultural tourism 7, 71, 132–133, 169, 185
　sustainability 9, 159–163
　tourism and agriculture relationship 11–12, 134–135, 138–142

trade connections with urban centres 35–36, 141
rural tourism 133, 134

safari tourism 117, 142, 170
San culture (Botswana) 114, 115
Shakaland cultural village (KwaZulu-Natal) 48, 171
Shona culture (Zimbabwe)
   dance performances for tourists 11, 122–123
   ethnic groups and languages 124
   traditional roles of music and dance 122
slavery, historical influence on carnivals 88
small to medium-sized enterprises (SMEs) 35, 95, 160
Social Exchange Theory (SET) 6
South Africa
   arts festivals 10, 80–83
   cultural diversity 66, 76
   national identity issues 9–10, 60, 65–67, 70
   political and administrative challenges 58–59, 68–72
   tourism policy objectives 3, 54, 65, 71
   Zulu heritage and cultural exchange 9, 49
   see also KwaZulu-Natal province
South African Tourism (SAT)
   data collection and reports 58, 60–61, 69
   strategic planning 18, 68–69
Southern Africa Development Community (SADC) 3
souvenirs
   authenticity 104–106
   sale during carnivals/events 95
   tourist demands 41
   video and photographs 124
Spatial Development Initiative (SDI) 3
special interest tourism (SIT) 59, 182
staged presentations
   authenticity and tourist acceptance 52, 103, 122, 129
   living museum model, cultural villages 148, 171
   reinforcing prejudice 6
statistical record keeping 58, 69, 95

storytelling
   role in indigenous knowledge preservation 37, 114
   used in Chapungu dance presentation 125–126
sustainability
   competitive threats for locally-owned ventures 52, 115–116
   elements, in tourism 41, 186
   ethnic/indigenous control of tourism 21–22, 27
   of livelihoods, critical analysis 12, 159–163
   of tourism, role of carnivals (Zimbabwe) 96, 97
Swaziland, Mantenga Cultural Village 26

telecommunications infrastructure 113, 141
Thaba-Bosiu cultural village (Lesotho) 40
thematic analysis of interview data 109, 122, 137–138
tourism industry
   recent trends towards flexibility 19, 59
   rise in cultural tourism importance 52–54, 165, 181, 184
   risks of unregulated development 117–118, 136, 185–186
   seasonal variation, value of carnivals 96
   tourist infrastructure development 112–113, 146
'tourist gaze' (Urry) 29, 51, 53, 124
traditional cultural heritage
   dress (clothing) 41, 104–105
   food 42, 140–141, 174, 175
   games 174
   herbal medicines 35, 42, 48, 175
   livelihood activities 108, 116
   music and dance traditions 122, 139, 175
   see also indigenous knowledge systems
Trinidad, carnival traditions 88–89
Tsiseb Conservancy (Namibia)
   community participation and power-sharing 163
   joint-venture accommodation partnership 152–154
   natural and cultural tourism resources 12, 147
   structure and area covered 146, 147

Tsodilo Hills (Botswana)
  archaeological importance and map 107–108
  benefits of commodification for tourism 113–114, 115–116
  interest/stakeholder groups 108–113
  local livelihoods 108, 112
  negative impacts of tourism development 10–11, 114–118

Ugab Wilderness Community Campsite (UWCC) 152–153
uKhahlamba-Drakensberg Park 39, 47, 48
understanding, cultural 6–7, 18, 38, 53, 82
UNESCO designation of World Heritage Sites 101, 183
United Nations Development Program (UNDP) 92–93

vandalism, at cultural sites 117, 118
village development committees (VDCs) 137
visitor experience
  expectations of World Heritage Sites 101
  interactions of history and culture 49, 169–170
  internet reviews 155
  management in wilderness areas 113, 170–171
  manipulation to emphasise Otherness 124–126
  quality, challenges in SA 65, 69
  value of effective site interpretation 151
visual arts 81–82
volunteer tourism 188

wages, lodge staff 156, 157, 160
warrior dance, Ndau people 127–128
water supply conflicts 116, 117, 118
White Lady paintings (Namibia) 145, 146, 152

White Paper on Development and Promotion of Tourism in South Africa (DEAT, 1996) 3, 66
wildlife
  human pressures on 117, 118
  management programmes 135
  photographic and hunting tourism 170, 171
  resources in southern Africa 3, 147, 183
wine festivals 80, 87
women
  roles and indigenous knowledge, Lesotho 9, 42
  traditional Basotho dress 41
  traditional food preparation and sale 140–141
World Heritage Sites (WHS)
  benefits of WHS status 54, 101, 115, 183
  Brandberg Mountain (Namibia) 145
  clustering and route development 71
  in KwaZulu-Natal (SA) 47, 48
  Maloti-Drakensberg (Lesotho/SA) 39
  stakeholders and management 102, 108–109
  Tsodilo Hills (Botswana) 107, 176
World Tourism Organisation (WTO)
  event hosted by Zimbabwe/Zambia 91
  predictions for cultural tourism 52–53

Zimbabwe
  objectives and launch of carnivals 89–90, 96, 97
  Shona music and dance heritage 11, 122
  see also Chapungu Dance Group; Harare international carnival
Zimbabwe Tourism Authority (ZTA) 89, 91, 94
Zulu culture (SA)
  emotional appeal for tourists 9, 51–52, 55
  historical heritage resources 48, 49

For Product Safety Concerns and Information please contact our EU Authorised Representative:

Easy Access System Europe

Mustamäe tee 50

10621 Tallinn

Estonia

gpsr.requests@easproject.com